501

Things <u>YOU</u>
Should Have
Learned
About...

MATH

METRO BOOKS
NEW YORK
An Imprint of Sterling Publishing
387 Part Avenue South
New York, NY 10016

METRO BOOKS and the distinctive Metro Books logo
are trademarks of Sterling Publishing Co., Inc.

Copyright 2014 Pulp Media Limited

This 2014 edition published by Metro Books
in arrangement with Pulp Media Limited.

All rights reserved. No part of this publication may be reproduced, stored
in a retrieval system or transmitted in
any form or by any means, electronic, mechanical,
photocopying, recording or otherwise, without the prior
written permission of the publishers and copyright holders.

AUTHOR: Sonia Mehta (in association with Quadrum Solutions)
SERIES ART DIRECTOR: Clare Barber
SERIES EDITOR: Helena Caldon
DESIGN & EDITING: Quadrum Solutions
PUBLISHER: James Tavendale

IMAGES courtesy of www.shutterstock.com;
www.istockphoto.com and www.clipart.com

ISBN 978-1-4351-4613-6

For information about custom editions, special
sales, and premium and corporate purchases,
please contact Sterling Special Sales at 800-805-5489
or specialsales@sterlingpublishing.com

Printed in China

10 9 8 7 6 5 4 3 2 1

www.sterlingpublishing.com

501

Things YOU Should Have Learned About...

MATH

METRO BOOKS
NEW YORK

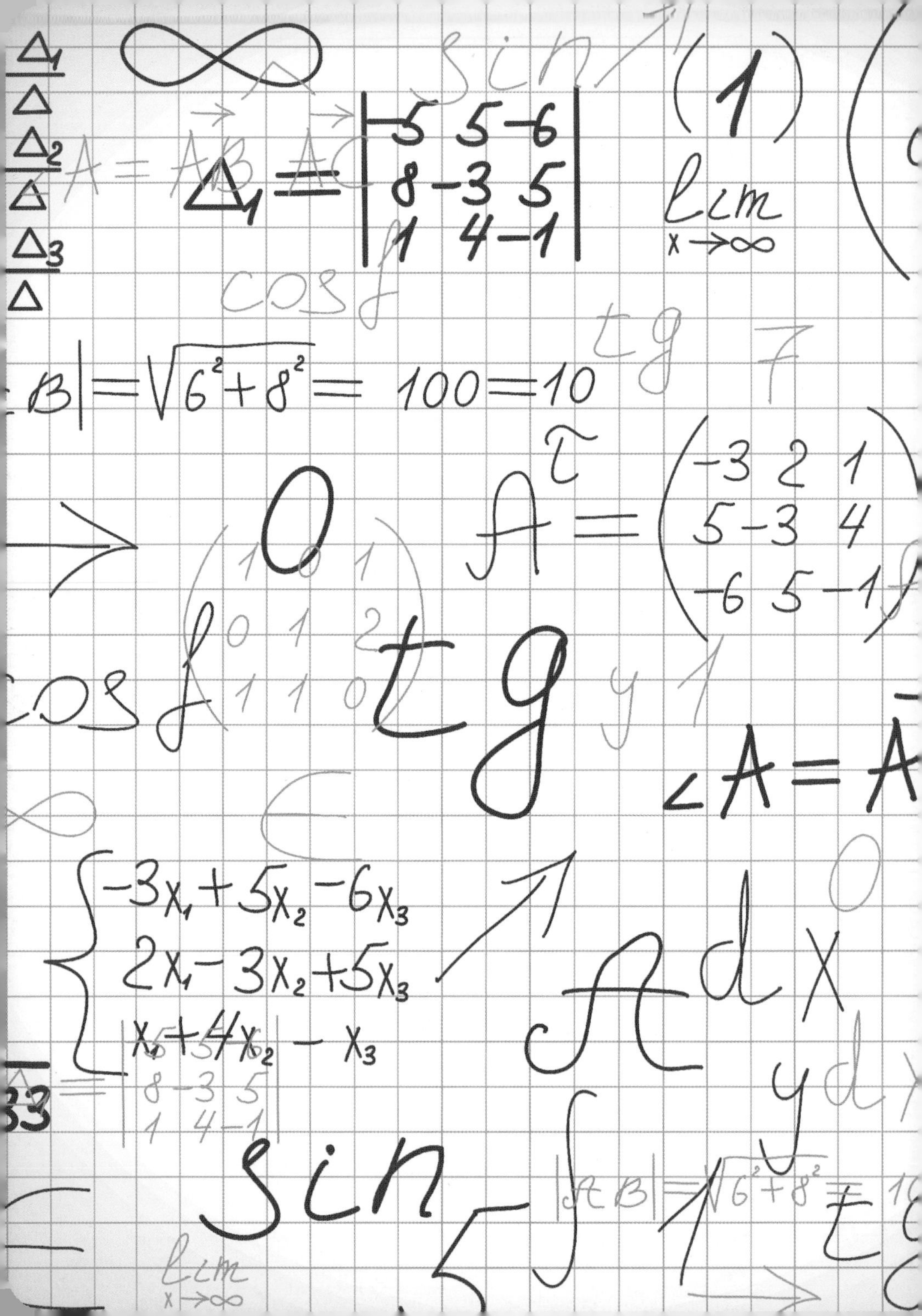

Δ_1
Δ_2
Δ_3

$\Delta_1 = \begin{vmatrix} 5 & 5 & -6 \\ 8 & -3 & 5 \\ 1 & 4 & -1 \end{vmatrix}$

$\lim\limits_{x \to \infty}$

$|B| = \sqrt{6^2 + 8^2} = \sqrt{100} = 10$

$A = \begin{pmatrix} -3 & 2 & 1 \\ 5 & -3 & 4 \\ -6 & 5 & -1 \end{pmatrix}$

$\angle A = \bar{A}$

$\begin{cases} -3x_1 + 5x_2 - 6x_3 \\ 2x_1 - 3x_2 + 5x_3 \\ x_1 + 4x_2 - x_3 \end{cases}$

$\int A\,dx$

$\int y\,dy$

$\sin$

$|AB| = \sqrt{6^2 + 8^2}$

$\lim\limits_{x \to \infty}$

CONTENTS

7	Introduction
8	Everything is in Number
38	Counting
50	Number systems, Tools and Calendars
82	Special Numbers and Shapes
112	Mathematicians, Lives, Deaths, Eccentricities and Superstitions
142	Maths and Culture
174	Math and Probability
204	Topology and Fermat's Theorem
214	Every Math
250	Index

INTRODUCTION

If you're one of those people, who believe Math is a dangerous territory to venture into, this book is for you. There are no complex, scary or horrific theorems in this book that will shake you to the bones; just facts about them that will definitely make you brighter and smarter. Therefore, the next time you attend a function with scientists and smart guys, make sure you browse through this book; it might just help you converse with them.

Each of the 501 facts included in this book has, in some way, contributed to the vast field of mathematics. Several facts in this book are bizarre, mind boggling, fun, and interesting, but not one will make you want to put it down.

Math is a science of numbers, an art of understanding shapes, a medium through which one can see the logic behind a hypothesis, and an aesthetic view of our world that is heavy dominated by numbers. This book will introduce you to the beginning of mathematics, right from the time when man was evolving until what's happening today. The book also explores mathematicians like Galileo Galilei, Archimedes, and Pythagoras, and their great contributions to the field of mathematics, their lives, deaths, and eccentricities. Through the book, you will get a chance to zoom across number systems, understand trivia, and figure out why mathematics is so useful, even though it may seem like a heavy discipline.

By the end of the book, your mind will be enriched, and you might even begin to see mathematics in a different light. Need to know more about Fermat's Last Theorem, Counting techniques, mathematics and astrology? Then what are you waiting for? Flip this page now!

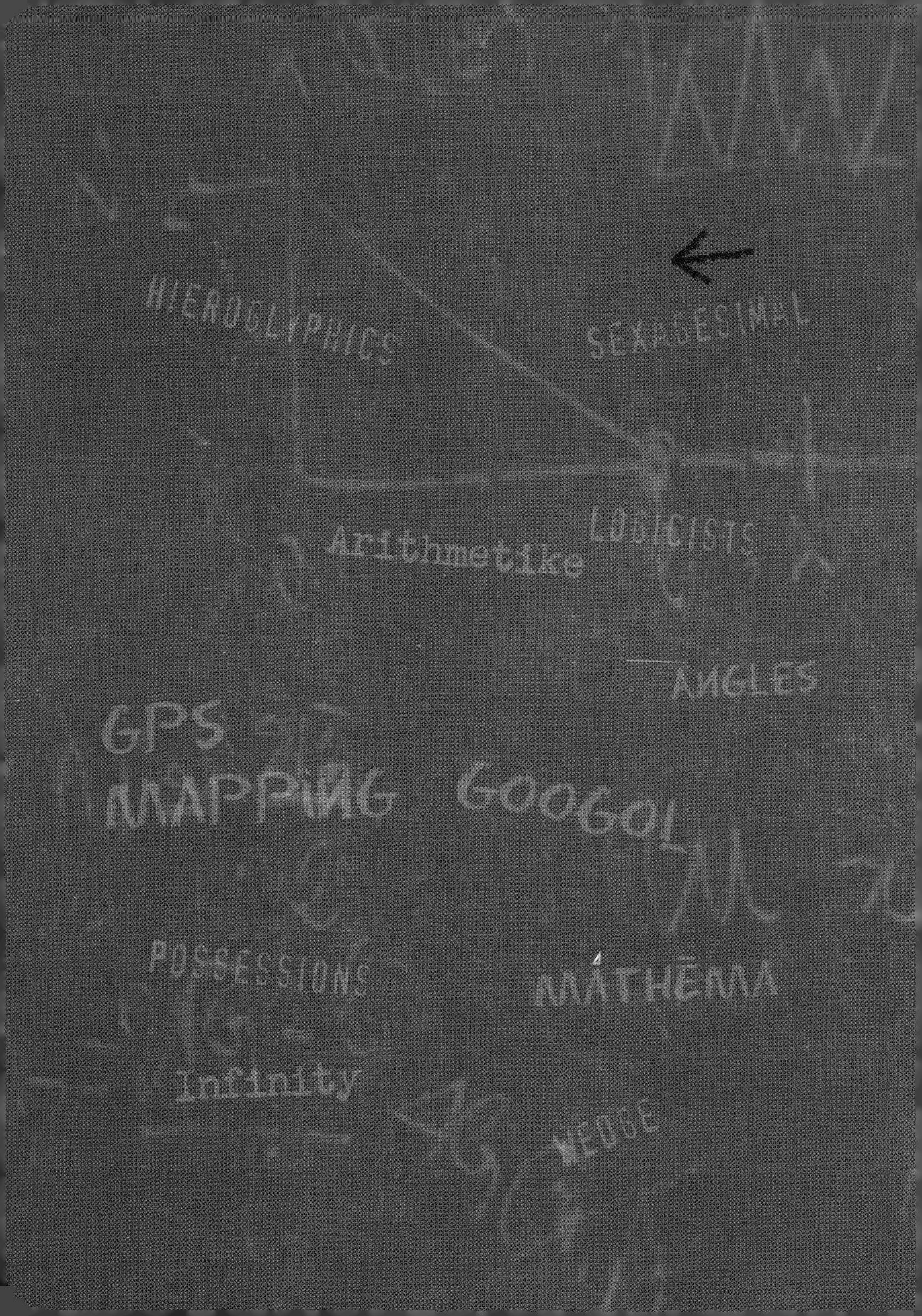

HIEROGLYPHICS   SEXAGESIMAL

Arithmetike   LOGICISTS

ANGLES

GPS
MAPPING   GOOGOL

POSSESSIONS   MATHEMA

Infinity   WEDGE

CUBES

501

MESOPOTAMIA

ROMAN

MATHEMATIKÓS

↓

Everything is a Number

BULLAS

ARITHMETIC

FRACTALS

KEY

Cuneiform

Mathematici

ZERO

1 THE WONDER OF MATHEMATICS

🎓 **MATHEMATICS IS A WONDROUS TOOL**. It is the key to the recognition of patterns around us that are discernible through any one or a combination of our senses. Man must have been so enchanted to see the consistency in the waxing and waning of the moon, the coming of the monsoon and floods, the periods of darkness and sunlight. His ears would have been delighted by the "music" produced by birds as they made sounds at intervals.

Man's search for truth turned him into a philosopher and mathematician. Mathematicians like Plato, Galileo and Pythagoras were philosophers and mystics at the same time. They attributed numbers with different qualities.

Galileo praised the magic of mathematics, claiming that the grand book of the Universe is written in the language of mathematics using geometric figures such as triangles and circles as characters.

Plato

2 FAST FACT

📖 **EVEN TILL** as late as the 18th century, mathematicians were considered to be eccentric people.

Galileo Galilei

Galileo said that without mathematics, one would stumble around in the darkness in one's search for the meaning of the universe. Plato believed that philosophers must know arithmetic (the theory of numbers) as they search for the truth. Bertrand Russell said that mathematics possesses not only truth but supreme beauty, which is sublimely pure and capable of stern perfection.

3 FAST FACT...

📖 **IN LATIN AND ENGLISH**, the term "mathematics" was originally used to denote astrology or, less frequently, astronomy. When St. Augustine, in the 1st century A.D., warned Christians to beware of "mathematici" they were being warned against astrologers and not mathematicians!

4 SCIENCE OR ART?

THROUGH CENTURIES, mathematicians, scientists, astronomers, architects, and artists have used mathematics to count, measure, calculate, deduce, beautify, predict, and prove. However, they still don't seem to agree on how to define it.

Certain mathematicians believe their vocation to be a science; others look upon it as an art form. The term "mathematics" is derived from the Greek word "máthēma", which in ancient Greek means "what one studies" or "what one knows".

Aristotle gave the first definition of mathematics, calling it the "science of quantity". The Sage dictionary defines mathematics as "the science dealing with the logic of quantity, shape, and arrangement". Is mathematics simply the manipulation of numbers using procedures like addition, division, etc.? Or is it, as Eugene Wigner describes it, "the science of skillful operations with concepts and rules invented just for this purpose?"

Pythagoras

Artists, and sculptors like Phidias looked upon mathematics as a tool of art and modeled their work on the ratios and proportions around them. But is it the job of mathematics to measure objects around us and turn aesthetics into numbers such as "phi"?

Opinions on this topic differ and though there are hundreds of definitions of mathematics, it is pretty safe to say that "mathematics has no definition" Let it suffice to say that "mathematics is what mathematicians do"; or as Pythagoras said, "Everything is a number."

5 FAST FACT...

IN AROUND 1920, H. L. Mencken jested about an unusual application of mathematics, saying, "It is now quite lawful for a Catholic woman to avoid pregnancy by resorting to mathematics, though she is still forbidden to resort to physics or chemistry".

6 FAST FACT...

THE WORD mathematics is also said to stem from the word "mathematikós" meaning "fond of learning".

7 BRANCHES OF MATHEMATICS

MATHEMATICAL STUDIES can be classified into two broad heads—pure and applied. Pure mathematicians regard mathematics as a thrilling mental exercise. They concern themselves with abstract issues like finding exactly how many prime numbers there may be, etc.

Pure mathematicians use logic to prove theorems built upon axioms. These theorems are true without doubt and exception, though they may have no practical utility in the physical world.

Applied mathematicians, on the other hand, try to understand and solve problems in the world using mathematical tools. They construct models based on existing problems in an attempt to find solutions for the problems. Pure and applied mathematics cannot be mutually exclusive.

Often, a theorem of pure mathematics comes to the aid of a problem which was unknown when the theorem was first proved. Pure mathematics then changes sides and becomes applied mathematics.

For example, in 1928, pure mathematicians set themselves the task to find out whether a procedure could be formulated where a mathematical statement could be judged to be true or false with the use of a finite number of steps. The solution of this abstract problem was used by applied mathematicians in the invention of the computer.

8 FAST FACT

MATHEMATICIANS may also be classified as logicists (people who use logic). Intuitionists are people who are concerned only with objects that can be constructed while Formalists only study symbols and formulae of mathematics.

Einstein

9 FAST FACT...

"PURE MATHEMATICS is, in its way, the poetry of logical ideas", said Albert Einstein.

10 KEEPING A RECORD OF POSSESSIONS

MAN SOON BEGAN TO REALIZE the need to keep a permanent record of his possessions that was uniform and unambiguous. Around 4000 B.C. man began to make tokens out of clay to use instead of stones, sticks, and bones.

In the hot and wet climate of Mesopotamia, clay was found in abundance, and when baked, it proved to be a good record. Thousands of tokens have been found which give us a picture of life in ancient times.

Tokens were made in geometric shapes of spheres, cones, disks, and rods they can even be thought of as a three dimensional writing system. The different shapes represented different things, or commodities. For example, a cone represented a measure of grain and an ovoid, a jar of oil.

To count different things man made different types of tokens. For instance, the token to count one goat differed from the one to count one sheep, and the one for ten goats differed from that for ten sheep.

Tokens represented both the object and its number, thus indicating that man had not yet understood the abstract concept of numbers.

Tokens

11 PROTECTION AGAINST FRAUD AND THEFT IN ANCIENT TIMES

THE CONCEPTS of ownership and commerce began to further safeguard man's interests. Advertently and inadvertently, the tokens representing records of payments, and goods possessed were likely to be broken or damaged. To prevent any dispute caused by ruptures or even loss and theft, man developed the idea of sealing tokens into envelopes made of clay. Fingerprints found on the inside of envelopes tell us that they were made by making a hole in a lump of clay using fingers. Often clay tokens were strung like beads on a string and then sealed to make them more tamper-proof.

Clay envelopes began to be marked on the outside to indicate the number of tokens held. Many envelopes have been found marked with impressions made by rolling a cylinder over their surface.

In order to prevent possible fraud by the cracking and re-sealing of envelopes, the markings on the outside of the envelope grew more complicated over time, so much so that any tampering would instantly become obvious.

12 FAST FACT

MARKINGS MADE by the people of Mesopotamia are the first written symbols for numbers, even though they cannot strictly be called writing.

In spite of these markings, if a dispute arose, the clay envelopes could always be broken open and the tokens counted.

The clay envelopes were called "bullas", a word derived from the Latin word "bullae". These tokens tell us that man was engaging in long distance commerce before he was writing.

Tokens and bullas are very important sources of information about the economic and social organization of those times as there is no written record for that time.

13 FAST FACT...

HISTORIAN DENISE Schmandt Bessaret's work of more than 20 years has given us knowledge about the meanings of the markings on the tokens and bullas.

14 THE INVENTION OF NUMBERS

MAN SOON ABANDONED THE BULLAS and made things simpler for himself by marking his "numbers" on tablets or slabs of clay. Most of the early tablets are records of what had been paid and to whom.

In the beginning man made his marks by pressing the relevant token the required number of times on to clay tablets. For example, to represent two jars of oil, he pressed the ovoid token twice on to the clay tablet.

Man soon began to turn to pictorial depictions and began representing things with pictograms. He drew jars, boars, etc., on clay tablets. Man realized that 10 sheep and 10 goats were of the same number. Man shortly grasped that there was no need to have separate symbols for 10 sheep and 10 goats.

Dissociation of the quantity of things from the type was something man began to pick up on. This was the beginning of the concept of abstract numbers.

Man started making signs for numbers independent of the object being counted, followed by a single pictogram for the thing being counted. This is the historic beginning of the understanding and writing of abstract numbers.

Historians have deduced that man has been writing numbers for 5,000 years now.

15 FAST FACT...

A TABLET FOUND inscribed with three circular imprints, three straight marks, and a pictogram for a jar of oil was interpreted as recording 33 jars of oil.

16 FAST FACT...

A MESOPOTAMIAN clay tablet recording beer rations for workers is on display at the British museum.

17 FAST FACT

GEORGE SAMUEL CLASON'S book "The Richest Man in Babylon" dispenses financial advice via an assembly of tales from ancient Babylon.

18 CUNEIFORM WRITING

📖 **IN THE BEGINNING, MAN** made number marks using curved and straight lines. He soon learnt to use a round stylus for the same. The number nine was written by pressing the stylus nine times. When pressed at different angles, the same stylus could be used to write different numbers.

Writing now became fast and uniform. Reeds being plentiful in Mesopotamia, man began using reeds as styluses to make wedge shaped marks. Such writing is known as cuneiform writing.

Cuneiform signs were pictographs. There was no need to know any language to read them. 1,900 cuneiform signs have been found on early clay tablets, including number and commodity signs.

Change from the earlier pictographic representation to the cuneiform representation also saw the orientation of the writing turning 90°. Writing became horizontal instead of vertical.

19 FAST FACT...
📖 **THE WORD CUNEIFORM** comes from the Latin term "cuneus" which means "wedge".

Many languages use the cuneiform script, including Sumerian, Akkadian, and Babylonian. Millions of clay tablets with cuneiform writing have survived the ravages of time and are displayed in several museums around the world.

20 FAST FACT
📖 **CUNEIFORM WRITING** was deciphered only in the 19th century.

21 USE OF BASE 10 BY SUMERIANS

🎓 **AS TOKENS STARTED BECOMING POPULAR**, man decided to go one step further and learnt to make marks on these tokens. This helped indicate the type and quantity of the objects being counted. The first sign man used was a simple cross and it indicated one sheep. Up to nine sheep were counted by an appropriate number of tokens.

22 FAST FACT...

📖 **THE EARLIEST CLAY** tokens have been discovered from Sumer (modern day Iraq and Iran), part of Mesopotamia, and are dated 4000 B.C.

To count 10 sheep, he made another type of mark. He represented 22 sheep by putting together two clay tokens which represented 10 sheep each, and two tokens which represented one sheep each. Man was thus adding even though he had no names for the numbers!

Tokens were made for agricultural and manufactured goods. Originally invented as record keepers, they soon began to be used for trade. Sumerians (inhabitants of southern Mesopotamia) made tokens for one object and ten objects, and counted these objects in terms of one or ten. This can be looked upon as the first use of a primitive base 10.

23 USING BASE 60 IN ANCIENT TIMES

🎓 **THE EARLIEST CUNEIFORM SYSTEM** of numbers used not just one but two bases—10 and 60. This means they used both the decimal and the sexagesimal system. They had unique marks for the numbers one, 10, and 60.

However, they did not have separate symbols for numbers up to 60. They made a suitable number of narrow wedges for the numbers up to nine. Wide wedges were used for 10. Multiples of 10 were marked by a suitable number of wide wedges.

Other numbers were represented by combinations of narrow and wide wedges. 60 and its multiples had separate symbols. 60 was a long vertical wedge, and 600 was written by marking a wide wedge mark of 10 on the vertical mark for 60.

3,600 was written by making a square like shape using four marks of 600.

Various numbers were written using combinations of these symbols. For example, 135 was written as a combination of two wedges denoting 60 each, one wedge denoting 10, and five wedges denoting one each.

How the ancient mathematicians came to use the sophisticated base of 60 is not understood. However, one conjecture is that it came into use as the number 60 has a large number of factors. This time however, the value of a numeral did not depend on its position in the number.

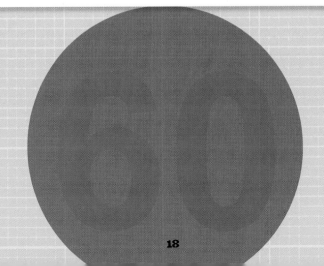

24 WHO INVENTED ARITHMETIC?

APART FROM SUMER in Africa (Mesopotamia), evidence of early writing has been found from other ancient civilizations of Egypt, Indus valley, China, and Central America. However, it cannot be said decisively if they all developed independently, and who influenced whom.

Sumerian numbers are assumed to have names, but we know of only two or three. Sumerians considered 3,600 as a large quantity. Both 3,600 and another "very" large number were represented by the word "Shar" in Sumerian.

Sumerians used the four operations of arithmetic, namely, addition, subtraction, division and multiplication. They had also devised symbols for multiplication, division, and for fractions.

25 FAST FACT...

THE NAME arithmetic comes from the Greek word "Arithmetike," meaning "the art of numbers."

26 FAST FACT

WE USE the sexagesimal system on a daily basis. For example, when we divide hours and minutes into 60 parts or say that a straight line makes an angle of 180 (60x3).

Modern man does not know the method by which man at that stage added and subtracted. This is probably because of the tradition of oral education. However, we do know how they multiplied. A mistake was made in a Sumerian multiplication table. Another tablet gave a corrected value. Comparison of the two revealed to modern mathematicians the Sumerians method of multiplication. The Sumerians are therefore honored as the inventors of arithmetic.

27 PLACE ACQUIRES VALUE

WHEN MAN WROTE numbers in ancient times; a numeral did not have a place value. That is to say, the value of a numeral did not depend on its position in the number. In modern systems, the 3 in 36 means thirty, and the 3 in 63 means three.

Ancient mathematicians soon realized that this was not good for trade and developed a place value system. Unlike before, there would be no confusion between 36 and 63. 36 would be written using three wide wedges followed by six small narrow wedges. 63 would be written using the long, narrow wedge for 60, followed by three small narrow wedges.

To achieve this degree of sophistication is one of the most remarkable achievements of that time. This invention allowed arithmetic to become sophisticated and the Sumerians learnt to add, subtract, divide, and multiply.

Tablets inscribed with multiplication tables for the numbers one to 20, 30, 40, 45, and 50 have been found. The Sumerians even learnt to find reciprocals, squares, and square roots of numbers as well as cubes and cube roots of numbers. Tables for these too have been found inscribed on tablets.

Tables for conversion of length, area, money, and weight have also been discovered. The concept of zero had not yet been envisaged. In fact, it would be thousands of years before zero began to be used universally.

x	0	1	2	3	4	5	6	7	8	9	10	11	12
0	0	0	0	0	0	0	0	0	0	0	0	0	0
1	0	1	2	3	4	5	6	7	8	9	10	11	12
2	0	2	4	6	8	10	12	14	16	18	20	22	24
3	0	3	6	9	12	15	18	21	24	27	30	33	36
4	0	4	8	12	16	20	24	28	32	36	40	44	48
5	0	5	10	15	20	25	30	35	40	45	50	55	60
6	0	6	12	18	24	30	36	42	48	54	60	66	72
7	0	7	14	21	28	35	42	49	56	63	70	77	84
8	0	8	16	24	32	40	48	56	64	72	80	88	96
9	0	9	18	27	36	45	54	63	72	81	90	99	108
10	0	10	20	30	40	50	60	70	80	90	100	110	120
11	0	11	22	33	44	55	66	77	88	99	110	121	132
12	0	12	24	36	48	60	72	84	96	108	120	132	144

28 FAST FACT

THE TABLETS with the multiplication tables were text books used by students as the teachers were probably unable to do complex calculations mentally.

29 BABYLONIANS AND THE PYTHAGORAS THEOREM

THOUGH THE PYTHAGORAS theorem ($a^2+b^2=c^2$) is named after the Greek mathematician of the same name, clay tablets from 1000 B.C. prove that the ancient Babylonians first realized the relationship between the sides of a right angled triangle and its hypotenuse.

In fact, "a", "b", and "c" in the equation are known as the Pythagorian triplets. When "a" and "b" are of value one, then c is the square root of two or 1.414213562 (correct to 10 decimal places) as per the Pythagorean theorem.

30 FAST FACT...

THE RATIO of the sides of size A papers used in modern times is constant and is equal to root two.

A clay tablet was found inscribed with a number equivalent to 1.414212956 (giving the value of root two correct to five decimal places) across the hypotenuse of a right angled triangle marked on it. This clay tablet is about eight centimeters in diameter. It is now at the Yale University as part of its Babylonian collection and is identified by the number YBC7289.

Many more clay tablets have also been discovered which prove that the Babylonians knew of the relationships between the sides of a right angled triangle.

A tablet known as the Plimpton 322 tablet has the Pythagorean triplets inscribed on it and also tells how to draw angles. The tablet is also inscribed with what may be thought of as a trigonometric table which seems to instruct on how to draw angles between 30 and 45 degrees.

31 FAST FACT

PARALLEL lines were used in the 16th A.D. to denote "equal to" which meant that no two things can be more equal.

32 HIEROGLYPHICS: THE EGYPTIAN NUMBERS

🎓 **THE EGYPTIANS**, in the 13th century B.C. used pictures to denote numbers. They drew pictures, called "hieroglyphs", to depict numbers. The number one was represented by a single vertical line, two by two lines, and so on till nine. Their system worked on a sort of base 10.

The Egyptians had pictograms for 10 and its multiples. 10 was represented by a circle; 100 by a coiled rope; 1,000 by a lotus blossom; 10,000 by a pointing finger and 1,00,000 by a tadpole.

The number 10,00,000 being mind-bogglingly large to the ancient Egyptians was depicted by a picture of a man with his arms spread wide in amazement. There was no zero in the system.

The writing of numbers in this system would look like illustrations in a present day story book for children. Thus, the number 2,00,300 would be represented by drawing two tadpoles and three coiled ropes.

The Egyptians had a symbol which looks like the zero but the symbol represented infinity, probably because a circle has no end. Thus, what we know as nothing, the Egyptians knew as infinite.

In spite of not having many symbols for numbers, the Egyptians were pretty advanced in their mathematical understanding; they knew how to represent fractions.

Egyptians had symbols for unit fraction, i.e., those fractions with the numerator one. They thus knew the importance of dividing one thing into parts. Apart from unit fractions, 2/3 and 3/4 were also important to the Egyptians and were given their own symbols.

33 DECIPHERING EGYPTIAN HIEROGLYPHICS

🎓 **TODAY, MAN HAS LEARNT** how to read Egyptian hieroglyphs thanks to the Rosetta stone on which the same text is written in two languages and three scripts. The two languages were Egyptian and Greek and the scripts were Demotic and ancient Greek.

Ancient tablet with heiroglyphs

Historians who learnt to decipher the ancient hieroglyphs found it extremely easy to compare what was written in an unknown script or language to what was written using a known one. This is how modern man learnt that a coil of rope or a tadpole was actually a number. The Rosetta stone was written in 196 B.C. and was discovered only in 1799 A.D. which is approximately only 200 years ago.

The Rhind papyrus was described by its writer as "direction for knowing all dark things" but it actually discusses mathematics. In the Rhind papyrus, different symbols were used to denote different mathematical operations. The symbol for addition was a pair of legs pointing forward. When the legs were pointing backward, it symbolized subtraction.

The Egyptian system of fractions was also revealed to modern man through the Rhind papyrus. The Egyptian system of fractions was based on "unit" fractions. The hieroglyph for ⅓ was a sort of oval with three small lines under it. That for ¼th had four lines under the oval.

34 FAST FACT

📖 **IN EGYPTIAN** hieroglyphics, subtraction was also symbolized by a flight of arrows!

35 FAST FACT

📖 **THE EGYPTIAN** system was not based on place value or orientation. Thus, the number 2,00,300 could be represented by two tadpoles and three coiled ropes written from left to right or from right to left or even vertically.

36 FAST FACT...

📖 **MIDDLE EGYPTIAN** history saw number names. Each number had a masculine and a feminine name.

37 FAST FACT...

📖 **HIEROGLYPHS** do not represent the objects themselves but the sounds that identify those objects.

38 THE ROMAN NUMBER SYSTEM

🎓 **THE ANCIENT ROMANS** had their own unique way of writing numbers. They used letters of the alphabet for this purpose. They used symbols for 10 and its multiples and for halves of these numbers. The numbers one, 10, 100, and 1,000 were represented as "I", "X", "C", and "M", while the numbers five, 50, and 500 were written as "V", "L", and "D" respectively. The Romans were aware of fractions and denoted half by the letter "S". This symbol was written before the integer. So, 10 ½ would be SX according to this system.

In the beginning, there was no place value in the Roman system. They wrote their numbers using both a "before" and "after" notation making it difficult to read numbers. So, 13 could be XIII or IIIX! 90 could be written as 10 less than 100 or XC. 7929 could be written as VIICMXXIX. Twisted and confusing, this can only be read by mentally spacing it out and reading it as (VII) (CM) (XX) (IX).

It is obvious that mathematical operations like simple addition and subtraction would be difficult in this system. Multiplication and other operations would be mind boggling. There was no zero in this system.

39 USING ROMAN NUMBERS

THE ROMAN WAY of writing numbers, though cumbersome, was used in Europe for nearly 2,000 years. A method was evolved to write bigger numbers with a limited number of numerals in the Roman script. These were small horizontal lines drawn above the letters.

The appendage signified multiplication by thousand. For example, X with an appendage meant 10,000. In modern times, sometimes numbers are written using the Roman script but not as the Romans would have written them. The Romans would have written MDCCCC for 1900 but in present times it is stylishly written as MCM in the "Roman" script.

Today, writing numbers in the Roman script is considered sophisticated. Clocks showing the numbers in Roman numbers are considered quite fashionable even in modern times. Many of them show IV as IIII. This practice comes from King Louis XIV of France who preferred to write four as IIII instead of IV. These clocks are known as Louis XIIII clocks. Roman numbers can also be seen on the facades of buildings.

They are used when issuing copyrights for films and TV shows. Successive Pentium processors launched by Intel were called Pentium I, Pentium II, Pentium III and so on.

They are also used in the names of wars, Popes and of British monarchs, for example, World War II, Pope Benedict XVI and Elizabeth II. The Olympiads in Beijing were called the XXIX Olympiads.

40 FAST FACT

THE SYMBOL V for five in the Roman number system comes from the shape formed by holding the thumb apart from the other fingers held close together; while X is a representation of the two hands crossed.

45 FAST FACT

📖 **THE ROMAN** numerals do not have the concept of zero, and the largest number that could be represented using this system is 4,999.

41 FAST FACT...

📖 **ROMAN NUMERALS** can only represent positive integers.

42 FAST FACT...

📖 **THE HEXAGESIMAL** system uses letters too for numbers and ABC in this system is 2748 in decimal.

43 FAST FACT...

📖 **ROMAN NUMERALS ARE** currently used quite often in law books in India.

44 FAST FACT

📖 **IN EUROPE**, Roman numbers were used traditionally to indicate the order in which children having the same names in a royal family were born. For example, the son of King Henry I was named King Henry II instead of King Henry Jr.

46 NUMBERS AND COUNTING IN ANCIENT CHINA

INDEPENDENT OF THE NUMBER SYSTEMS in the west, a number system was developed and used by the Chinese. Archeological discoveries in 1899 in Xiao dun in China brought to light the advance system of accounting in China.

Bones and tortoise shells were discovered on which numbers were written. They recorded the number of war casualties, the number of animals hunted. There were even records which could be seen as calendars.

Symbols for the numbers were in the form of straight and curved lines. The numbers were written using a decimal system. They even had positional values. They had separate symbols for the numbers one to nine and for 10, 100, and 1,000. Other numbers were written as combinations of these numbers. The largest number found on this excavation site was 30,000.

The Chinese knew of the arithmetical operations. There is also evidence of the Chinese using fractions. The shells and bones found in Xiao dun dates back to around 14th century B.C. If this is correct, it would mean that the Chinese number was more ancient than the Sumerian system.

47 FAST FACT

THE CHINESE number system uses ideograms and is called "ideographic" writing. Ideograms are pictures of what is to be depicted.

48 THE MAYAN NUMBER SYSTEM

THE MAYAN CIVILIZATION FLOURISHED in and around Mexico around the 1st century B.C. It evolved independently from the other ancient civilizations. They were such great mathematicians that they constructed architectural wonders without the use of metal tools or even the use of the wheel.
The ruins of Chichen Itza are testimony to their skill.

They developed their own number system of notation with place value. The mathematical system of the Mayans was unique in that they used the base 20. This is known as a vigesimal system. The numbers were not placed from left to right like our system but vertically, from top to bottom. They used a place holder for zero. Its symbol was in the shape of a sea shell.

The Mayans used combinations of dots and bars to write their numbers. Each dot represented the numeral one and each bar, five. They were able to represent huge numbers using only these two symbols.

Mayans could represent numbers using stones, pebbles, bean, bones, sticks, etc. They were able to represent extremely large numbers using just three symbols and wrote dates which took up several lines to write. The Mayans did not have a broad written language. So, most of their knowledge is lost.
They used knotted strings called Quipu as record keepers.

49 FAST FACT...

THE MAYAN pyramid at Chichen Itza was used as a calendar. Its steps add up to 365, the number of days in a year.

The Mayans are also credited for inventing the calendar, whose modified form we use today. Their calendar kept track of time in three ways. The Long Count keeps track of all days since the calendar began. The Tzolkin is a divine calendar. The Haab keeps track of the year.

50 FAST FACT

THE MAYANS calculated that the number of days in a year is slightly more than 365 using just sticks and without any knowledge of fractions.

51 THE ANCIENT INDIAN NUMBER SYSTEM

ANCIENT INDIAN MATHEMATICIANS gave the world the decimal number system which is most commonly used today. It is known as the Hindu-Arabic number system and it originated in India. Modified forms of the written representation of Indian (Hindu) numerals one to nine and zero are the numerals used most commonly the world over in current times.

The zero was also invented by ancient Indian mathematicians, though which mathematician used it first is a debated issue.

Ancient Indian books, the Vedas, are collections of hymns or mantras in praise of the creator. These were handed down orally by the ancient sages as early as 1,000 B.C. and were compiled into books much later.

Mathematical rules and problems were set out in verse in ancient India as this made it possible to state them with the least possible jargon and they came with the ease of recall. In these hymns, figures up to a trillion are mentioned. The four arithmetical operations are also spoken of.

Powers of ten, fractions, multiplication, squares, and cube roots were well-known to ancient Indians. It is claimed that in the 4th century. Indians discussed numbers the equivalent of 10 to the power of 421 which is much larger than the estimated number of atoms in the universe. The Indian text "Sulba Sutra" speaks of the Pythagorean triplets.

Ancient Indians experimented with algebra and trigonometry. A book called "Surya Sidhanta" whose authorship is unknown; contains text dealing with sine, cosine, angles, etc.

52 FAST FACT

IN 1100 A.D. Indian mathematician Bhaskara used the zero in Algebra.

Indian mathematicians from Kerala evolved a series of expansions for trigonometric functions; creating the first known power series. This was done nearly two centuries before calculus was invented in Europe. Aryabhata, a well-known Indian mathematician, wrote of quadratic equations, the value of pi and perfected the decimal system.

53 CALCULATING WITHOUT THE ZERO

IT IS ASTOUNDING to learn that the pyramids were built, astronomical calculations were carried out, time was measured, square and cube roots were found, fractions were written—all without the number zero!

None of the great ancient mathematicians like the Sumerians, Egyptians or even Greeks, realized the importance of a zero quantity.

55 FAST FACT...

IT HAS BEEN SPECULATED that the Sumerians did not miss having a zero because the large number factors of 60 made it possible to represent most numbers with ease.

Even when positional number systems were used by the Babylonians, nearly 3,000 years since the first clay tokens were made, there was no symbol for zero. In this system they left an empty space to denote "nothing".

The practice made a difference between 123 and 1,203 clear. However, 1,230 and 123 could still be confusing since leaving an empty space at the end of a number is also considered meaningless (today, however, the zero only after a decimal point is considered meaningless).

The positions of the stars were calculated without knowing whether a number was 123 or 1,230 (written in the sexagesimal system). Values were inferred from the context. The empty space at the end was later filled in by marks such as two slanted wedges. These marks were never used at the end of quantities like 1,230. They can be thought of as punctuation marks rather than numbers.

54 FAST FACT

THE SYMBOL of two slanted wedges to symbolize nothing was first used by the Babylonians in around 300 B.C.

The world changed and progressed and the number zero was yet to be discovered!

56 THE CONFUSION OF THE GREEKS

THE ANCIENT GREEKS in spite of the immense advances made by them in the realm of mathematics, were confused about the concept of nothingness. The Greeks asked themselves, how could nothing be something?

Philosophers and religious leaders had endless debates about how this nothingness or vacuum could exist and what could be its nature. Many paradoxes, unsolved for thousands of years, are attributed to Greek mathematician Zeno of Elea. He put forth three premises, the first of which was that "any unit has magnitude".

Zeno of Elea further argued that "if something when added to or subtracted from something else, fails to change it, then that something is nothing". This gives us a glimpse of the early understanding of nothing or zero.

However, they did not deduce from these thoughts that zero could be a number. Aristotle thought of both infinity and zero as ideas rather than actual numbers.

58 FAST FACT

THOUGH ANCIENT GREEKS like Archimedes guessed that enormous quantities existed, such as the number of grains of sand on a beach, without the convenience of a zero, it was impossible to write them down.

Ἀρχιμήδης
Archimedes

57 FAST FACT...

IN AROUND 130 A.D. Claudius Ptolemy, a Greek mathematician, used the symbol of a small circle with a bar as a place holder for nothing, but this was only in the fractional part of the number.

59 THE INVENTION OF THE ZERO IN INDIA

INDIAN MATHEMATICIAN BRAHMAGUPTA, who lived in the 7th century A.D., was the first to conceptualize zero as a number and write out elaborate rules for it. The credit however, is given to Indian mathematician Aryabhata who lived in the 5th century A.D.

Georges Ifrah, a French mathematician, was one of the first to give credit of inventing the zero to Aryabhata. His calculations of square and cube roots could not be possible if he did not recognize the concept of zero in the place value system.

Zero, in Aryabhata's system was the place holder for the power of 10 with a valueless co-efficient. He used the word "kha" to denote emptiness zero. His statement, "sthanamsthanam das gunam" means "place to place is ten times in value and is the basis of the place value decimal system".

60 FAST FACT...

"THE IMPORTANCE of the creation of the zero mark can never be exaggerated. This giving to airy nothing, not merely a local habitation and a name, a picture, a symbol but helpful power, is the characteristic of the Hindu race whence it sprang." said G.B.Halsted, a famous mathematician.

61 FAST FACT

A JAIN TEXT, the Lokavibhâga, from around the 5th century A.D. is believed to be the oldest existing text in the world to mention a decimal place-value system which includes a zero.

62 FAST FACT...

A SMALL CIRCLE, the universally recognized symbol for zero and used as such, can be seen inscribed on a stone at the Chaturbhuja Temple at Gwalior in India. It is the oldest inscription of the zero known to mankind.

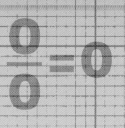

63. THE FIRST RULES FOR CALCULATING USING THE NUMBER ZERO

THE ZERO MADE ITS grand entry in the work of Brahmagupta, titled the Brahmasputha Siddhanta, and dates back to 628 A.D. Brahmasputha Siddhanta translates as "Correctly Established Doctrine of Brahma" and is also known as "The Opening of the Universe". This is the first written record of rules for working with zero as a number.

In this book, Siddhanta set out explicit rules for the use of zero or "shunya". He gave rules, such as the sum of zero and a positive number is positive, and the sum of zero and a negative number is negative. He also stated that the sum of zero and zero is zero. The problem of division and multiplication by zero which had confused Greek mathematicians was dealt mathematically by him.

Siddhanta also mentioned that if a number is divided by zero, it can be thought of as a fraction with zero as the denominator, and vice versa. He ruled that zero divided by zero is zero.

Modern mathematicians, however, do not assign a value to this number believing this number to be indeterminate or infinite. Brahmasputha Siddhanta also gave rules for the use of negative numbers, for computing square roots, for equations, for summing series, etc.

64. FAST FACT...

THE BOOK
Brahmasputha Siddhanta is written completely in verse.

65 FROM INDIA TO THE WORLD THROUGH ARABIA

🎓 **ANCIENT INDIAN MATHEMATICAL KNOWLEDGE** was carried to Arabia via travelers. Calculations using these methods involved moving numbers around and even erasing them. Due to this requirement Arabian mathematicians wrote on a "dust board". Indian numerals came to be known as "ghubar" (the Arabian word for dust) numerals because of the dust generated in these calculations.

Information about these numerals and the wonderful invention of zero by the Hindus of India was written about by Al-Khwarizmi, a Persian mathematician, in his book, "On the Calculation with Hindu Numerals" in about 825 A.D. The original Arabic text of On the Calculation with Hindu Numerals is lost to humanity.

In addition, Al-Khwarizmi wrote his own explanation of the use of zero in the same book. Thus, he brought the Indian system of writing numbers in which numerals had place values and the zero into Islamic civilization. His work was translated as late as the 12th century into Latin in the book "Algoritmi de numero Indorum". The name means "Al-Khwarizmi" or "the Hindu Art aof Reckoning".

The oldest surviving book about the Indian system written in Arabia is the "Kitab al-fusul fi as-hisab al-Hindi". It was written in about the middle of the 10th century by "Al-Uqlidisi".

The system was brought to the notice of the western world by mathematician Fibonacci in his book "Liber Abaci", which was published in 1202. "Liber Abaci" means "the book of counting". It was only then that the western world became aware of the wealth of ancient Indian mathematics.

Since the knowledge of zero came to them via the Arabic alphabet, they named the system describe in Fibonacci's book, the Hindu-Arabic number system. The zero took about 600 years to travel from the East to the West.

Even after this publication, the use of these numerals became widespread in Europe only in the 15th century A.D.

Centuries have seen exponential advances in mathematics, none of which would have been possible without the zero. Today, the Hindu-Arabic system is the most commonly used number system in the world.

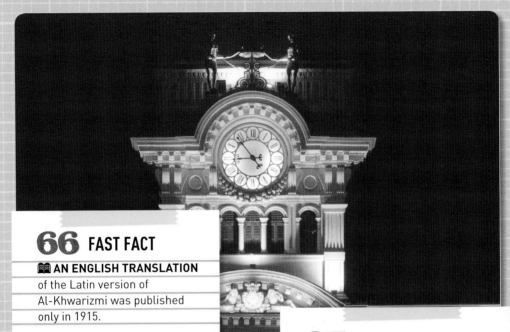

66 FAST FACT

📖 **AN ENGLISH TRANSLATION** of the Latin version of Al-Khwarizmi was published only in 1915.

67 FAST FACT...

📖 **THE NAME 'ALGEBRA'** is derived from the term "al-jabr", which is an operation used by Al-Khwarizmi to solve quadratic equations.

68 FAST FACT...

📖 **THE LATIN FORM** of Al-Khwarizmi's name, "Algoritmi", is the root for the names 'algorism' and 'algorithm'.

69 FAST FACT...

📖 **IN THE BEGINNING**, the government of Italy outlawed the use of Arabic numbers including zero. However, merchants found it so useful that they used it illegally.

70 THE MANY NAMES OF 'NOTHING'

THE NUMBER THAT HAS THE LARGEST variety of names is zero. Some say zero comes from Zephirum, which means empty.

Cipher is another name originating from the Arab word "sifr". Nought or naught is an old name for zero in American English. It comes from the old English word "nowiht".

Zero is also understood as nil and this comes from the Latin word "nihil". Probably the newest English name for zero is the letter O (often spelled oh). This is used when giving telephone numbers and addresses. It is also used when talking about time, as in four o' five which means five minutes past four o'clock.

In tennis, zero is called "love". In cricket, it is called "duck". Some others names for zero are zip, zilch, nix and ought.

"Nada" which is a Spanish word is another name for zero, and introduced into the English language by an author called Ernest Hemingway.

71 FAST FACT...

THE NAME O or Oh is used in the designation of the popular Hollywood spy James Bond. He is popularly known as 007 (double O seven).

72 FAST FACT...

ABSOLUTE ZERO (-459.69°F/-273°C) is the lowest temperature theoretically. It may never be reached.

73 FAST FACT...

THE NAME "cipher" means a message written in code or a secret method of writing as well as zero. It also means a person with no influence.

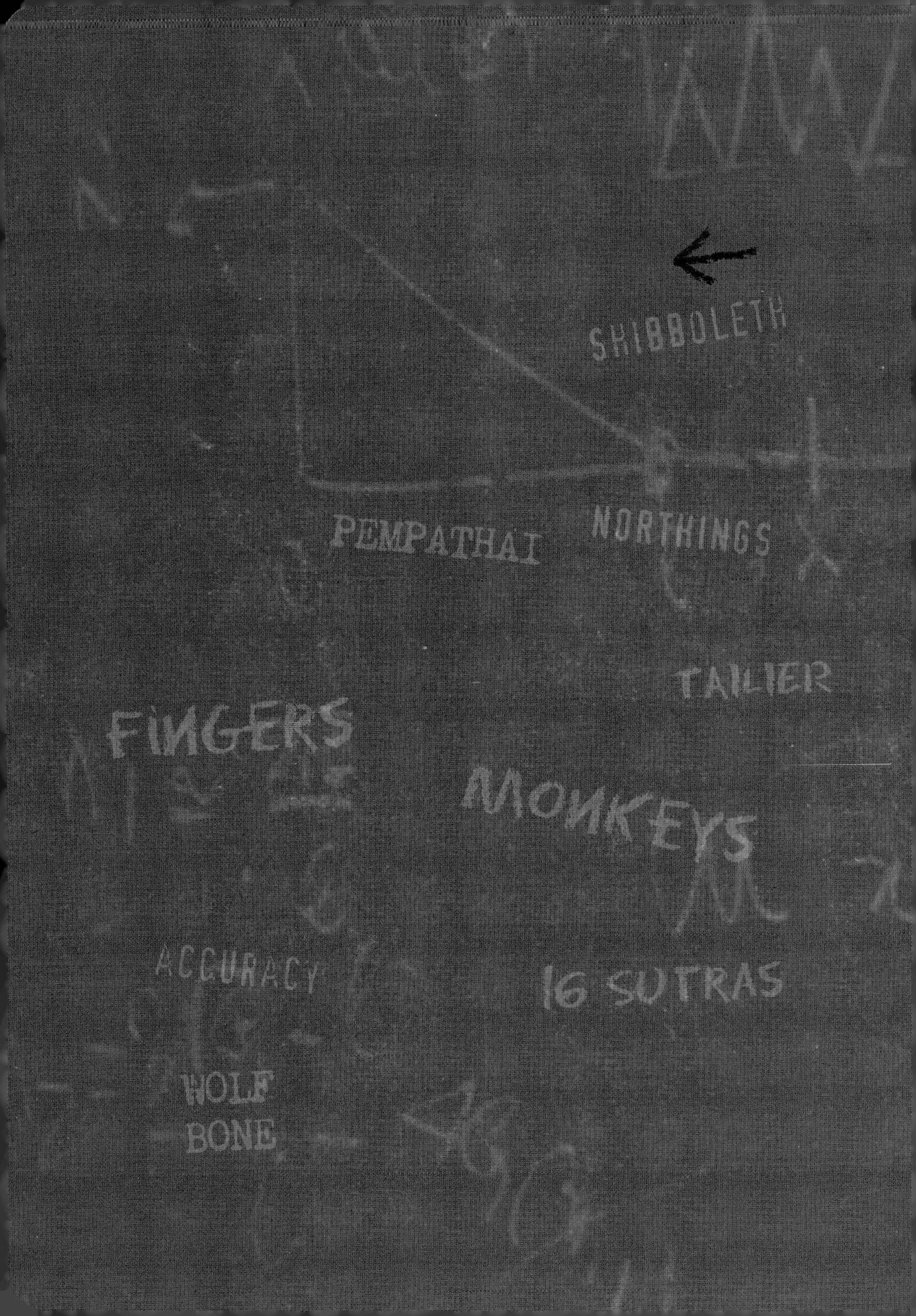

SHIBBOLETH ←

PEMPATHAI   NORTHINGS

TAILIER

FINGERS

MONKEYS

ACCURACY   16 SUTRAS

WOLF
BONE

501 Counting

DACTYLONOMY
VEDA
MARKING
CHANTING 
ISHANGO BONE
LUNAR
JAP
TALLY

74 HOW MANY DO I HAVE?

📖 **MATHEMATICIANS AND HISTORIANS** believe that the first mathematical concept understood by primitive man was "counting". As man began to amass possessions, he must have felt the need to know how many he owned. The fingers of his two hands being the most readily available and uniform counting tools, man must have first used his fingers to count on.

In the times of Homer, the Greek word for counting was "pempathai", which means counting by fives. This suggests that they counted on their five fingers.

Homer

English historian, Bede in his book, De temporumratione, describes a system that allows one to calculate up to 9,999 using the fingers on his two hands. The lifting up of two fingers would indicate, two things being counted.

Other readily available counting tools would probably have been stones, sticks and bones. Man may have kept track of things, possibly via a one-to-one correspondence. Two fruits could have been indicated by putting two stones side by side, and so on. The larger the pile of stones, the richer the owner was considered to be (using the "greater than" and "less than" concept).

75 FAST FACT...

📖 **IN THE BOOK OF THE DEAD**, an old Egyptian book from around 50 A.D., it is suggested that in order to obtain a ferry, one must be able to number one's fingers.

76 FAST FACT

📖 **THE ANCIENT** Egyptian symbol for 10,000 is a single raised finger.

Primitive man learnt to count before he learnt to make symbols, or "write". He may have counted while speaking the language, which itself was rudimentary without any names for numbers.

Early man did not have the concept of "nothing". On being asked how many apples he had left after eating the one he had, he would have been very confused.

77 COUNTING FINGERS DIFFERENTLY

PEOPLE FROM DIFFERENT CULTURES raise different fingers to represent the same number. The English raise their index finger for the number one, and add one finger for each successive number till four. Five is indicated by raising all fingers and the thumb of any one hand. Six is represented by the digits of one hand along with the index finger of the other.

The western Europeans, however, extend the thumb to represent one. In some cultures, this means go ahead. Western Europeans indicate two by extending their thumb and index finger, and so on.

Japanese people count with their fingers in one way when counting for oneself and in another when indicating numbers to others. When counting for oneself, five is represented by a closed palm.

When signaling to others, the Japanese follow the English system the number five. Numbers greater than five are shown by holding the appropriate number of fingers against the open palm of the other hand. However, 10 is shown with the fingers of both hands extended.

Cultural differences like these are termed "shibboleth" and used to identify different cultures instantly and inadvertently.

78 FAST FACT

📖 **IN THE MOVIE** the Inglourious Basterds, Lieutenant Archie Hicox, a Jewish-American posing as a German, gave himself away when he held up the 'wrong' three fingers to signal the number three.

79 FAST FACT...

📖 **THE STUDY** of finger counting is known as "dactylonomy".

80 CAN ANIMALS DO MATHEMATICS?

🎓 **IN AN EXPERIMENT** conducted with olive baboons, it was found that their instinctive mathematical understanding equaled to that of human toddlers. The University of Rochester reported that researchers placed 1 to 8 peanuts in two cups and allowed captive Olive baboons to choose a cup.

The subjects were allowed to take the peanuts from the cup they chose. It was seen that in 75% of 54 trials, the subjects chose the cup having more, when the number of peanuts in the cups differed by a mere three. The study found that the results were similar to the cognitive ability of toddlers who had not been taught to count. Therefore, the concept of "greater than" and "less than" seems to be natural to both man and baboons.

Rhesus monkeys were found to be faster than college students in choosing the set having a smaller number of objects out of two given sets. It must be admitted, however, that their accuracy was lower.

Rhesus monkeys proved that they could even use more than one of their senses. In an experiment conducted at Duke University, the monkeys could match the number of sounds they heard to the number of shapes they saw.

81 FAST FACT...

📖 **EVEN ROBINS** understand the concept of greater than and less than. In a study conducted in New Zealand they invariably gathered around the holes in which they had seen more worms being put.

The domestic chicken has also been found to be aware of these concepts. The birds were placed in front of two opaque screens. One ball was sent behind the first screen and four behind the other. The bird followed the four balls. It has been proved that tiny insects like ants can measure distance by "counting" the steps taken.

To test if ants actually counted steps, scientists Muller and Wehner made stilts for the ants using pig hair. This made their steps longer and the ants miscalculated. It is certain that they have an instinctive sense of geometry which they use to build their anthills.

82 FAST FACT

📖 **THE VENUS** fly trap is a plant that knows how to count and even understands the concept of time. It does not snap shut as soon as something lands on it. However, if another hair on its leaf is touched within 20 seconds of the first touch, it snap shut!

83 COUNTING CHANTING

🎓 **THE FIVE FINGERS OF EACH** hand were used by early man to indicate numbers from one to five. Using the fingers of the other hand too allowed the representation of numbers up to 10.

In some civilizations, man learnt to count larger numbers using finger joints or phalanges. Ancient Egyptians used the joints of four fingers on one hand to count forward, using the thumb to count with. They counted up to 12 on one hand.

A similar method of counting is used by the Hindus of India to perform a "jap" even today. Jap refers to chanting mantras or repeating the name of a Hindu God very softly or silently in the mind. It is done using the 12 phalanges on the right hand to keep track of the number of chants, while using the thumb as a pointer.

The first jap is done with the thumb touching the top of the little finger of the same hand. The second is done with the thumb pointing to the middle phalange. Counting on all the phalanges makes twelve japs. This is one round.

The thumb of the left hand is positioned on the tip of the little finger of the left hand to mark this. In this system, there is no need to know the names of numbers to "count" and thus allows even illiterate people keep track of the number of "rounds" of japs done.

Since the number 108 is considered holy by the Hindus; they usually do nine rounds of 12 japs each to make 108. This is an example of an ancient practical use of the base 12 by non-mathematicians.

84 A FAST FACT...

📖 **PEOPLE FROM SEVERAL** cultures (Jews, Christians, etc.) use prayer beads on strings to keep count of their holy chanting.

85 MARKING 'NUMBERS'

SINCE FINGERS, toes, sticks, and stones were not very effective long-term record keepers, primitive man learnt to cut notches and make markings to indicate quantity.

Man began using these instruments to keep a record as early as the Paleolithic times, i.e., more than 35,000 years ago. Not having invented numbers, he made marks on bones as a one-to-one correspondence basis.

The marks observed are called tally marks and may be interpreted as the earliest written signs for numbers. It is natural that early man would use bones for this purpose. Bones with such markings on them are referred to as tally sticks. The word tally comes from the French word "tailier" meaning to cut. Early man however, did not have a name for them.

Bones have been found, some very recently, with markings on them. The oldest is believed to be more than 30,000 years old. It is the bone of a wolf and is simply called the "Wolf bone". The Wolf bone is marked with 55 marks, grouped in sets of five. This could be interpreted as the use of base five.

Other ancient tally sticks that were found are the "Ishango bone" (more than 20,000 years old) and the "Lebombo" bone. Both are baboon fibulas and were discovered in Africa. The Lebombo bone is 7.7 cm long and is marked with 29 notches. The markings on the Ishango bone are grouped in three columns running down its length.

86 WHAT DO THE 'TALLY' MARKS MEAN?

🎓 **SCIENTISTS GENERALLY DISAGREE** as to what purpose these markings could have served. Most mathematicians and historians believe that they are simple tally marks to facilitate counting.

Certain others state that the grouping of these markings or notches indicate a mathematical understanding that seems very sophisticated. This has led some mathematicians to suggest that the markings may even indicate a number system.

Other scientists suggest that early man may have made the markings to keep track of the lunar cycles.

It is also hypothesized that the markings on the Ishango bone might represent a six-month lunar calendar. Mathematician Peter Rudman believes that the marks may represent prime numbers.

It has also been proposed that both the Ishango bone and the Lebombo bone may have been marked to keep track of the menstrual cycle. The menstrual and lunar cycle being similar, the markings can be interpreted to mark either or both.

On the other hand, some scientists state that the markings merely provide a better grip as these tools were most probably used for engraving, seeing that the Ishango bone does have a sharp piece of quartz attached to one end which could have been used for engraving.

87 FAST FACT

📖 **AS THE MARKINGS** on the Ishango bone correspond to the menstrual cycle of females, historian Claudia Zaslavsky believes that it was marked by a woman.

88 FAST FACT...

📖 **BUSHMEN IN NAMIBIA** continue to use calendar sticks that are similar to the Lebombo bone.

89 FAST FACT...

📖 **THE WOLF BONE WAS** discovered only as late as 1937 in Czechoslovakia. Till then the Lebombo bone was considered by some as the world's oldest artifact.

90 ANCIENT INDIAN SYSTEM OF QUICK CALCULATION

🎓 **VEDAS ARE ANCIENT INDIAN TEXTS** or books of knowledge belonging to the period around 1500 B.C. to 500 B.C.

The Sanskrit word "Veda" means "to know without limit". One of these Vedas contains certain aphorisms or word formulae (sutras) which were carefully reconstructed by an Indian mathematician called Swami Bharati Krishna Tirtha.

A set of 16 sutras form the base of the system of Vedic Math. These sutras are very simple to use and make numerical calculations easy and quicker than the conventional math system.

91 FAST FACT

📖 **THE ATHARVAVEDA** is the religious text which contains all the 16 sutras of Vedic Math.

In Vedic Math students can arrive at the correct solution to a math problem in many different ways. These techniques give a person liberty to choose the method that he/she is most comfortable with.

Methods of Vedic Math provide an insight into all streams of math and build a strong foundation and understanding of concepts like logic, trigonometry, calculus, etc. Fractions and multiplication tables become simple once the steps of Vedic Math are applied.

92 RELEVANCE OF VEDIC MATH IN MODERN TIMES

THE VEDIC MATH SYSTEM is characterized by coherence and interrelation, making the system very easy to grasp, even for small children.

People who find math tedious, find this system to be playful and stress free. This stress free quality ensures that all fear of math is eradicated and students look forward to solving complicated problems.

With practice, a person improves his /her speed and numerical skills. This method is said to induce analytical thinking and a bold approach to solving a math problem.

It is useful for students, professionals, teachers, and parents. Students find this system of calculation beneficial during exams because it helps them save time in big calculations. There is scope for fewer errors and gives better results.

93 FAST FACT

SWAMI BHARATI KRISHNA Tirtha wrote 16 volumes of Vedic math in from 1911 to 1918, but all of them were lost and he had to rewrite them from memory in 1956.

BINARY

hexadecimal   OCTAL

                                    DECIMAL

ANCIENT
INDIAN   SHUNYA

OBELUS   ADDITION

EQUAL TO

501 ↓ Number systems, tools and calendars

94 THE BINARY NUMBER SYSTEM

THE BINARY NUMBER SYSTEM comprises of only two symbols, zero and one. It is generally used in computers and several other digital electronic devises. The binary system is however, not a recent finding. This system was invented almost three centuries ago!

Gottfried Wilhelm Leibniz, one of the inventors of calculus, published a paper called "Essay d' une nouvelle science des nombres", which contained his work on the binary system. He had written this paper for his election into the Paris Academy. However, the actual discovery of this system happened almost 20 years before his work.

The computer memory recognizes two elements, one and zero, also known as a "bit", which is short for "binary" digit. A set of eight bits is called a "byte". These are used in computers as a series of "on" and "off" switches, where each digit's place value is twice as much as the digit on its right.

Gottfried Wilhelm Leibniz

The binary system is also called Machine Code, because it is used in different electronic machines, ranging from small machines like simple calculators to complex structures like super computers.

This code is considered to be the base level of software, as all other software needs to be converted into machine code before it can be used in computer programs.

95 FAST FACT...

📖 **THE HEXADECIMAL SYSTEM** uses 16 digits to represent real numbers. The digits used are 0, 1, 2, 3, 4, 5, 6, 7, 8, 9, A, B, C, D, E, and F. It is a system used in computer programing, where four bits can be represented as one hexadecimal digit.

96 FAST FACT...

📖 **AS THE NAME SUGGESTS**, the Octal System uses eight digits to represent real numbers. Did you know that Tom Lehrer wrote a song called "New Math" (That Was The Year That Was, 1965)? The song explains how to compute 342-173 in Octal!

97 THE DECIMAL SYSTEM

🎓 **IT WAS THE HINDUS** who devised a method of expressing numbers with the help of the decimal system. The Hindu mathematicians used a base 10 system. Having invented the zero or "shunya" it became easier to write big numbers.

98 FAST FACT

📖 **THE OPERATING SYSTEM** on your desktop or laptop comes with a calculator which converts from decimal to binary system and the other way round too. There are several tools available online to help you do this simple conversion as well.

According to famous French mathematician Pierre Simon Laplace, "It is India that gave us the ingenious method of expressing all numbers by means of 10 symbols, each symbol receiving a value of position as well as an absolute value; a profound and important idea which appears so simple to us now that we ignore its true merit. But its very simplicity and the great ease with which it has lent to computations has put our arithmetic in the first rank of useful inventions; and we shall appreciate the grandeur of the achievement even more when we remember that it escaped the genius of Archimedes and Apollonius, two of the greatest men produced by antiquity"

Pierre Simon Laplace

The advantage of this system lies in the fact that only 10 symbols (1, 2, 3, 4, 5, 6, 7, 8, 9, and 0) can be used to write values, ranging from the tiniest of fractions to the largest of quantities.

A decimal point is used to display the whole number on the left and the fractional part on the right. For example, in the number 16.3334, 16 is the whole number and 3334 is the fractional part.

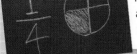

99 FAST FACT...

📖 **WE USE FRACTIONS** everyday in our daily lives without even realizing it. We use measuring spoons in the kitchen for cooking our favorite recipes, giving the correct dosage of a medicine to a sick person, cutting a pizza in slices, using money in several denominations, etc.

100 MATHEMATICAL SYMBOLS

🎓 **GALILEO SAID,** "Mathematics is the language with which God wrote the Universe." Math is considered a language because it has its own rules and script, just like other languages around the world.

The symbols used in mathematics have their own history. The basic fundamental operations like addition and subtraction got their symbols in the 14th and 15th centuries. Here are some fascinating facts on these symbols:

Nicole d' Oresme, a Frenchman used the "+" symbol in his work for the purpose of addition. In the year 1489, Johannes Widmann's "Mercantile Arithmetic" had plus and minus signs in print form.

The division symbol "(÷)" was called Obelus and was used in 1659, while William Oughtred was credited with introducing the cross symbol used to denote multiplication. Three British mathematicians Harriot, Oughtred, and Barrow are known to have developed symbols for "greater than" and "lesser than".

Johann Gauss was the person who gave us a symbol to denote congruency. The symbol to denote factorial has its roots in Switzerland, Germany, and France.

In 1557, Robert Recorde created an important mathematical symbol of "equal to"; while in 1655, John Wallis devised the symbol for infinity.

These symbols of mathematics have become an integral part of our lives. People from around the world understand the language of mathematics easily, due to the simplicity of these symbols.

101 FAST FACT...

📖 **AN IMPORTANT ASPECT** of the number system is working with calculations involving negative numbers. A real number, which is less than zero, is called a negative number. Negative numbers are used to signify a loss, deficit or decrease in a particular quantity. These numbers are denoted by the minus (-) sign.

102 FAST FACT...

📖 **THE RELATIONSHIP** between negative and positive numbers is very easy to understand with the help of a number line. The numbers appearing on the right of zero are positive real numbers and the ones appearing on the left of zero are negative numbers.

103 FAST FACT...

📖 **AN IMAGINARY** number is any number whose square is a negative number. A complex number is any number which can be expressed in the form of "a+bi" where "a" and "b" are real numbers and "i" is the imaginary number. They are usually denoted by "i". These numbers are used in the creation of complex numbers, which are used in many scientific applications like the quantum theory, cartography, etc.

104 FAST FACT...

📖 **REPRESENTING COMPLEX** numbers on a graph enables us to understand and analyze them better. These are called Argand diagrams, and are named after amateur mathematician Jean Robert Argand who gave the idea of representing complex numbers in a geometrical way.

105 USING FINGERS AS CALCULATING TOOLS

EARLY MAN USED fingers to count on. Thus, fingers were the earliest mathematical tools. With time, man must have performed arithmetical operations of adding and subtracting using his fingers. However, fingers can be and are used for easy multiplication too.

For example, to multiply by 9, the fingers are numbered 1 to 10 starting at either thumb. Let us begin at the left thumb. To multiply by say, 4, the finger numbered 4 is bent down. Then, the fingers to the left of the bent finger (3) give the number to be placed in the tens place. The number in the units place is given by the numbers to the right of the bent finger (6). Thus, without remembering the tables of nine, one can multiply using fingers.

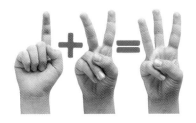

In order to multiply by any number from 6 to 10, there is another easy way.
The fingers of both hands are labeled from 6 to 10 starting at the small finger.

Multiplying any two numbers, say, six and seven, the finger numbered six of one hand is touched to the finger number seven of the other hand. Then, the number of fingers including and below the two touching fingers on each hand are counted.

When we multiply the sum of these two numbers by 10; we have the value 30. In the next step, the number of fingers above the joining fingers are counted and multiplied. We get the value, 12 (4x3=12). Adding 30 and 12, we get 42 which is the product of six and seven!

106 FAST FACT

THE EGYPTIANS multiplied and divided by repeated addition and subtraction.

107 THE ABACUS – A CHINESE CALCULATING TOOL

THE ABACUS WAS DEVELOPED in China around 3000 B.C. It is a tool which helps facilitate mathematical calculations.

The beads are used in an abacus to do calculations. The abacus does not do the calculations like computer. The word abacus comes from the word "ibq". This is a word from the Semitic language and means "to wipe the dust".

An abacus consists of a frame, usually wooden; there are a number of parallel wires. There is a bar perpendicular to the wires dividing the wires into two sections.

On each wire, there are two beads above the bar and five beads below the bar. The two beads represent five units each. The five beads on the other side of the bar represent one unit each. The wire on the extreme right represents the units place. The one to its left represents the tens place and so on. Numbers are represented by shifting the appropriate beads on the bar dividing the wire.

108 FAST FACT...

Chinese counting boards were developed in around 1000 B.C. The abacus was used by Greeks and Romans around 500 B.C.

📖 **PEOPLE WHO ARE** proficient at using an abacus claim that they can calculate faster than a computer by using an abacus.

109 NAPIER'S BONES AND LOGARITHM TABLES

IT WAS NOT POSSIBLE to calculate large numbers using primitive methods. Around 1614, John Napier devised a system of calculation using the powers of numbers to reduce this difficulty.

Napier based his system on the fact that numbers can be multiplied simply by adding their powers. For example, the product of 2^3, and 2^4 is simply equal to 2^7. He wrote out logarithm tables which gave the value of the powers of numbers at a glance. This made it easy to multiply large numbers.

Based on the same principle, Napier devised a set also known as rods. These rods were marked with numbers. They were called Napier's bones or Napier's rods. He wrote about this idea in a book called "Rabdologia" in 1617.

110 FAST FACT...

PROFESSIONAL MATHEMATICIANS in China made use of counting rods along with the abacus to perform mathematical calculations for accounting and financial transactions.

Some people call this method of calculation "Rabdology". Using the product of any number and a single digit number, could be worked out much faster than before. However, it would seem cumbersome to use as it involved a lot of calculation.

A system called "Genaille rods" was developed soon afterwards in which the products could be directly read without any need for calculation.

111 PASCALINES

🎓 **IN THE YEAR 1642,** Blaise Pascal invented the first mechanical calculator in France. Pascal was a child prodigy. He did not have any formal education and was home schooled by his father. To help his father who was a tax collector he devised the calculator to help with laborious tax and accounting. He called it the Pascaline.

Blaise Pascal

Pascal continued to improve on the first model and built around 50 versions of these Pascalines. The early calculator was in the shape of a brass box with wheels inside it. These wheels were turned with a stylus.

The calculator could perform all four mathematical calculations, i.e., addition, subtraction, division, and multiplication. The answers would show up on top of the calculator. This was the first time in history that these operations were performed by the machine itself.

The machine performed all mathematical operations in terms of addition. It subtracted by adding the complement of the number to be subtracted to the first number. Multiplication was done by repeated addition. Division was in terms of subtraction.

112 FAST FACT...

📖 **ANOTHER NAME** for Pascal's calculator was Arithmetique.

113 FAST FACT...

📖 **PASCAL TOOK** 20 years to complete the first logarithmic tables.

114 FAST FACT

📖 **IN 1616,** Pascal suggested the use of the decimal point in the writing of numbers in such a way that the digits to the right of the decimal represented the fractional part of the number.

As the machine was developed in France, the answers appeared in terms of deniers, sous, and livres. However, the calculators could be tweaked to calculate in other currencies too. Although, a revolutionary idea for that time; using these calculators was still quite laborious.

115 INTRODUCTION OF COMPUTERS

MECHANICAL CALCULATORS, made calculations fast, but were not perfect. There were a lot of errors in the computations. In 1823, Charles Babbage, a British engineer, built a steam powered machine which was capable of automatically calculating logarithms. He called it the "Difference engine".

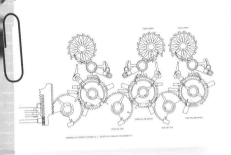

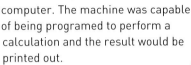

The "Difference Engine" is generally considered a precursor to the modern electronic computer. The machine was capable of being programed to perform a calculation and the result would be printed out.

However, Babbage was not able to get enough money to sustain and improve his machine and had to abandon the project before he could perfect his machine.

Leibniz improved on Pascal's machine and invented a calculating machine called the "Step Reckoner". This machine was based on the decimal system even though Leibniz perfected the binary system. The binary system is used by computers even today.

The first generation of computers was built in the 1940s and used vacuum tubes. The second generation was built in the 1950s and made use of transistors. The third generation of computers was introduced around the 1960s and had integrated circuits.

Electronic calculators today are inexpensive, fast, and capable of performing scientific as well as simple arithmetical calculations. They were commercially available only in the 1960s.

116 FAST FACT

THE COMPUTER program ADA is named after Ada Lovelace, daughter of Lord Byron, who wrote the programs for Babbage's difference engine.

117 FAST FACT...

IBM INTRODUCED the first electronic calculator in 1948.

118 FAST FACT...

JAPAN NEC developed the first electronic computer in 1958.

119 NATURE AND THE MEASUREMENT OF TIME

THE EARLY EGYPTIANS were intrigued by a group of constellations, called "Decans". Every year, when the Nile was about to flood, the Egyptians observed that 12 Decans appeared in the sky. This made the number 12 important to them.

The Egyptians devised a calendar which was divided into 12 months. Each interval of daylight in the calendar was divided into 12 hours. The Egyptians did not think of the hours of darkness and light as part of a single day.

Instead, they divided the periods of darkness independently into 12 divisions too. However, as the duration of the Sun varied each day, the Egyptian hours could not be fixed and varied from day to day, and season to season.

In 150 B.C. Hipparchus, a great mathematician observed that periods with sunlight and darkness were of equal duration on days of the equinox every year. Based on this, he proposed summing up the periods of darkness and daylight into 24 equal hours. However, his idea did not find favor because it did not hold true for the other days.

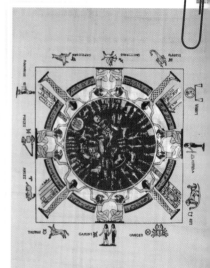

Egyptian Calendar

120 FAST FACT...

EGYPTIAN SUNDIALS, which were stakes pushed into the ground, are the first instruments to have been used by man to keep track of hours.

121 FAST FACT...

Greek Mathematician **ERATOSTHENES**, in 194 B.C., measured the angle of the shadow to the Earth using vertical rods in two places to estimate the circumference of the Earth.

122 TIME RUNS OUT

THE EGYPTIANS DEVISED an ingenious method to tell time during the hours of darkness. They invented a fascinating device called the "Clepsydra" or "water clock".

The name Clepsydra is made up of two Greek words, "klepto" and "hydro" meaning "thief" and "water". Therefore, the word actually means water-thief. This was an apt name, as the Clepsydra was a simple container filled with water whose bottom had a hole in it. The container was calibrated with 12 divisions along the sides. The water escaped through the hole and the level gradually decreased. The Egyptians knew the time by looking at the calibrations at which the level of water rested.

In 600 B.C., another instrument called the "Merkhet" was developed by the Egyptians to tell time. Two Merkhets were placed to mark a line running from north to south. The Egyptians could "tell" the time by observing when specific stars crossed over this line.

The Clepsydra, however, was the most efficient means of telling time in that period.

123 FAST FACT...

IN AROUND 1088, a mechanical clock driven by escaping water was developed in the Far East, which not only told time but also gave astrological and astronomical information.

124 MEASURING TIME ACCURATELY

🎓 **TELLING TIME** has come a long way since the Egyptians. Several other civilizations like the Greek, Roman, etc. worked with hours of differing durations for nearly 2,500 years. It was only after mechanical clocks came into use during the 13th century that people accepted the idea of dividing the day into 24 hours.

Early clocks were based on the principle of the verge and foliot mechanism, but they were not accurate enough. In 1656, Christian Huygens made the first pendulum clock, but Galileo is often credited for it. Pendulum clocks were quite accurate, erroneous about only one minute in a day.

In the 1930s, Quartz clocks came into being. Quartz crystals used in these timepieces vibrate about 100,000 times per second. These are so accurate that the error was limited to only one second in a month.

125 FAST FACT...

📖 **STANDARD TIME** is agreed to be time on an imaginary line, which runs from north to south in Greenwich, England. The line is used as a comparison to measure time all over the world. For every degree east or west of this time zone, the change varies by an hour.

126 FAST FACT

📖 **CLOCKS ARE** advanced in the summer to make better use of the hours of sunlight in the evening. This concept is called "Daylight Savings Time".

127 FAST FACT

📖 **IN THE SOLAR** observatory at Kodaikanal, India, there is a pendulum clock which has been telling accurate time for over 200 years without being serviced.

Galileo Galilei

128 STANDARDIZING DISTANCES IN ANCIENT TIMES

🎓 **THE CONCEPTS OF LARGE AND SMALL,** tall and short are all relative. In order to construct and compare, everyone must have a standard to refer to.

In ancient times, measurements of length were in terms of human body parts. Measurements for quantities were in terms of fingers. Measurements for distances were in terms of strides.

The units used for measurements like strides, hands, and feet varied from human to human, and yet remarkable architectural wonders have been created. During the construction of the pyramids, the cubit was used as a unit of measurement.

A cubit is the measure from the elbow to the tip of the middle finger of an adult human and it varies between 15 and 20 inches. Pyramids could not have been built without the standardization of the cubit. The cubit was standardized using the black granite kept by those in power.

Official measuring sticks used in the construction of the pyramids had to confirm to this. Using these rudimentary measuring instruments, the Egyptians built the Great Pyramid of Khufu, each of whose sides measure 1050 feet in length and is 480 feet in height. The sides had to be built at correct angles or else they would not meet at the top.

The cubit was one example of a body part used as a measure; here are some more interesting ones:

The breadth and length of a single finger was also used in ancient times to measure small distance. The length of a "foot" was called a foot and used as a unit by the Romans.

The measure from the tip of the nose when a person is looking straight ahead to the thumb of his outstretched arm is one yard or 3 feet. The same with the head turned to the opposite side is considered to be 3.2 feet. However, body parts are rarely used to measure today as they can only give approximate measurements.

129 FAST FACT

📖 **ROMANS DEFINED** a mile to be equal to the distance covered in 100 paces.

130 FAST FACT

📖 **THE DISTANCE** between the Sun and the Earth was worded out in ancient times by measuring the distance between two cities. The paces between them were counted along with the angle of shadows cast by two sundials belonging to the two cities.

131 STANDARDIZING WEIGHTS IN ANCIENT TIMES

THE BODY WAS ALSO used as a standard for weight in ancient times. In India, for example, when emperors were pleased, they gave the favored person gifts equal to their weight. The person was made to sit on one side of a weighing scale and was weighed against grain or gold, which was then gifted to that person.

Another way to show their generosity was when the king was weighed and the goods were distributed amongst the poor. It did pay to have an overweight king!

In order to facilitate and flourish their trading practices; the Sumerians developed standard weights made from stone or metal. These weighed around 640 to 978 gms and called "minas". 60 minas made a "biltu". Smaller units were called "siqlu". Metal coins of fixed weight developed from these.

132 FAST FACT

STANDARD WEIGHTS were not only made in the usual round or square shapes but also in the shape of elephants (in Burma) or warriors (in Africa).

Standardization of weights was attempted in China in 221 B.C. Vessels for measuring grains were standardized. The ancient Chinese understood the relationship between notes produced when a metal object is struck and its weight; so the standard vessels for weighing were of equal weight and made the same note when struck. Thus, if the shape, material, and weight were common, vessels of the same volume would make the same sound. In ancient Chinese, the words for wine bowl, bell, and grain measure were the same.

In India too, when coins were made of gold and silver, the purity of the metal was determined by the sound it made when tossed or flicked by a thumb nail.

133. MEASURING DISTANCES IN CURRENT TIMES

THE CGPM ("Conference Generale des Poids et Merures" or the "General Conference on Weights and Measures") was held in 1875. This was an intergovernmental treaty in which nearly 50 countries agreed to standardize units and use the SI system of units. This was done in the interest of smooth advancement of science and technology.

The metric system is proving to be an internationally workable and accepted system of units. This has the advantage of the decimal system, and conversions and notations are easy.

Originally a meter was defined by this organization to be one ten-millionth of the distance from the North Pole to the equator on the meridian running through Paris.

The first prototype or standard meter was made from a platinum-iridium alloy in 1874. It was shorter than the meter by 0.02 mms because of the flattening of the Earth caused by its rotation.

In 1889, the use of a new alloy reduced this error to less than 0.0001 to be measured at the melting point of ice. In 1927, a more precise standard was adopted. The meter was defined as the distance between two marked points on a platinum-iridium bar kept at BIPM. This was to be measured at a temperature of 0° and standard atmospheric pressure. The bar was to be supported by two cylinders kept at a distance of 571 mms from each other. Subsequently, even more stringent definitions were given. Now, the meter is defined as the distance traveled by light in a vacuum at 1/299792458 seconds.

134. FAST FACT

THE BAR WHICH was the prototype of 1889 is still kept in France under the same conditions as in 1889.

135. FAST FACT

THE DECIMAL POINT was introduced into the decimal system only in 1616 A.D. by John Napier.

136 UNCOMMON UNITS OF DISTANCE

🎓 **OLIVER R SMOOT** used his height to measure the Harvard Bridge in 1958. This was meant to be a prank. A plaque to commemorate the occasion records the measure to be 364.4 "smoots" plus or minus an ear.

137 FAST FACT

📖 **THE MIT COMMUNITY** Running Club measures distances in smoots.

138 FAST FACT

📖 **THE EXPRESSION** "rule of thumb" has its roots in practice of carpenters of taking rough measurements using their thumbs.

139 FAST FACT

📖 **THE UNIT "DIRAC"** was coined to measure the level of talkativeness of the quiet Paul Dirac who was a theoretical physicist. One Dirac was one word per hour.

Another uncommon unit in actual use is the "barn". Its roots lie in folks saying "could not hit the broad side of a barn". The difficulty of getting particles to collide in particle accelerators is very high.

Although, the size of barns is not quite small, a barn was defined as a unit to measure really tiny areas like the cross section of atomic nuclei. A barn is equal to 10^{28} square meters. Other units of similar nature are even smaller, such as the "outhouse" and the "shed". However, these units are rarely used.

A pun on light years is the unit "beard second". It is defined as the amount a beard grows in one second. It is equal to 100 angstroms. Donkey power is defined as one-third of horse power. It is approximately 250 watts.

140 STANDARD UNITS OF WEIGHTS IN MODERN TIMES

THE BRITISH IMPERIAL SYSTEM of weights originates from the middle ages, when it was called "avoirdupois". This word comes from the French word "aveir de peis" which means "weight of goods". The system was used widely for hundreds of years.

It was then introduced officially by the Magna Carta in 1215.

In this system, the conversion of units does not follow any particular pattern. A mile is 1760 yards, a yard is 3 feet and 1 foot is 12 inches. The problem is compounded by the fact that the measures are not standardized internationally and there are slight variations in between countries.

The metric standard of mass is the platinum-iridium prototype made in 1889 and kept at the International Bureau of Weights and Measures. It was declared in the 3rd CGP 1901, that the kilogram is the unit of mass and is equal to the mass of the international prototype of the kilogram.

Popularly, the weight of any object is calculated in terms of kilograms, grams, etc., but the weight of an object is dependent on gravity. Thus, the unit of weight in SI units is Newton. For example, an object of mass 2 pounds has a weight equal to 9.8 newtons on the surface of the Earth.

141 FAST FACT

THE WEIGHT OF COINS made it difficult for merchants across the world to carry large amounts on their journeys. Merchants left their coins in safe keeping. The receipts they carried for their coins led to the idea of paper money.

142 MEASURING AREA

🎓 **ANCIENT ROMANS BUILT** a number of public buildings and amenities. They made detailed maps of the towns based on the surveyors' measurements. They divided the land into squares called "centuriae". Accurate squares were actually marked out on the land as well as on the map.

In modern times, computers are used to determine the optimum usage of area. For example, in a readymade garment industry, computers determine the least amount (area) of cloth from which the pieces required for a garment can be cut. Then, laser beams are used to cut along those directions.

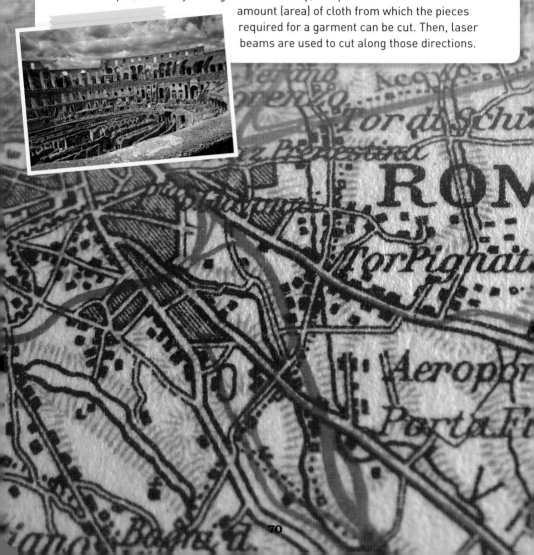

143 STAYING AFLOAT

🎓 **THE VOLUME OF** an irregular object can be measured by using the volume of liquid displaced by the object. You might know about Archimedes and his experiences with the displacement of water!

Archimedes' laws hold true, because when ships are loaded with cargo, heavier loads would cause a greater displacement of water and the ships would sink lower. If the ship is too heavy, it might sink so much that it becomes unstable.

Samuel Plimsoll, a British merchant, thought of marking lines on the sides of ships to indicate the level at which the ship would remain stable. His idea was adopted internationally in 1930, and these lines are now called International Load Lines.

Hypothetically, if we were to put Earth and Saturn in pools of water, Saturn would stay afloat but Earth would sink! This is because though the mass of Saturn is 100 times that of Earth, its relative density is only 0.7 and that of Earth is 5.5.

144 FAST FACT

📖 **THE FORMULA FOR** the volume of a sphere ($4/3\ \pi r^3$) is inscribed on Archimedes' tombstone.

145 MAYAN MATHEMATICIANS AND THE CALENDAR

THE MAYANS WERE ACCOMPLISHED mathematicians and followed a precise and elaborate calendar system, which comprised of three separate systems integrated in a tricky format. They merged a 260 day calendar with a 365 day one, where only one day of the two calendars would coincide once every 52 years!

The 260 day calendar was called the Tzolkin, and the 365 day calendar was called the Haab. The Mayan astronomers used extensive mathematical calculations to merge the dates of these two calendars.

The third calendar system that the Mayans followed was called the Long Count which was just a count or a time scale. Each event was marked on it like a scale. The Mayan mathematical system was called the Vigesimal system and developed on a base of 20, unlike the decimal system which is on a base of 10. Historians discovered that dates on the ancient Mayan inscriptions had to be multiplied by 20 to translate them into modern dates.

All the Mayan calendar calculations were done keeping in mind the cycles of celestial bodies like the Sun, Moon, Venus, etc. This calendar was later adopted by the Aztecs who changed the names of the months while keeping the mathematical calculations the same as the Mayan calendar.

146 FAST FACT

THE MAYAN PYRAMID in Mexico is actually an enormous calendar. It was built in the year 1050. It has 91 steps on each side, leading to a large platform on the top. All the steps add up to total 365, signifying one day a year.

147 FAST FACT

📖 **MAYAN MATHEMATICIANS** could conduct detailed astronomy calculations with remarkable ease and precision. They used gadgets like two sticks mounted in the form of a cross for viewing celestial bodies through the angle that the sticks formed. The Caracol building in Chichen Itza is considered to be an ancient observatory because many of the windows were carefully positioned for viewing equinoxes and other celestial movements.

148 FAST FACT

📖 **THE FIVE UNLUCKY** days of the Haab calendar were called "Wayeb". The Mayans did not wash or comb their hair or do any hard work during this period. People who were born on these days were considered unlucky and destined to live in poverty for their entire lives!

149 MATHEMATICS AND THE BABYLONIAN CALENDAR

THE ANCIENT BABYLONIANS, like the Mayans, were very accomplished in math. They used a base 60 system called the "Sexagesimal System". Even today we use this system to count hours, minutes, and seconds.

Unlike our measurements of 10 years being called a "decade" or 100 years being called a "century", the Babylonians called a period of 60 years a "Soss", a period of 600 years a "Ner", and a period of 3,600 years a "Sar".

The Spring Equinox was the beginning of a year for the Babylonians. There were 12 months in a year based on the lunar cycles and 354 days leading to a problem of a shortfall of days.

The problem was later corrected by the Egyptians who added six days, making it a total of 360. The Egyptian King Ptolemy was credited with adding a single day at the end of four years bringing in the concept of the Leap Year.

150 FAST FACT

THE BABYLONIANS used a wedge and line system to form numbers and letters. This form of writing is called the cuneiform. The famous code of Hammurabi is written in the cuneiform format.

151 FAST FACT

THE BABYLONIANS devised an ancient Clepsydra which is a type of wheel for measuring time. The Clepsydra is circular and divided into 60 hours, minutes, and seconds. It can be traced back to Babylonian era.

152 FAST FACT

THE ANCIENT BABYLONIAN sundial was a flat disc with markings. The object that casts a shadow on the flat surface is called a Gnomon. Sundials usually show hours from sunrise to sunset.

153 MATHEMATICS AND THE JULIAN CALENDAR

JULIUS CAESAR WAS responsible for modifying the inaccuracies of the Roman calendar. In 45 B.C., a new calendar was declared by him. He called it the Julian calendar. This calendar was based on two calculations; the solar year (the time between two equinoxes) and the sidereal year (the time taken for the Earth to complete one orbit around the Sun).

Caesar's mathematicians found that even after incorporating the leap day in February every four years, there was a small error which eventually resulted in a lack of synchronization in the seasons.

Better gadgets were soon devised for calculations and it was discovered that even an error of a single day in a 100 year cycle became glaringly apparent over the coming centuries. The Julian calendar got ahead by one day every 125 years!

In order to correct this discrepancy, in the 1500s Pope Gregory XIII suggested that some days in a month could be cancelled to bring the calendar back on track. This was met with much opposition from the Protestants and they rejected a change of such magnitude for many years.

The change was finally incorporated after 200 years and in the month of September 1752, September 2, was to be followed by September 14.

154 THE GREGORIAN CALENDAR AND MATHEMATICS

MOST COUNTRIES HAD adopted the decree given by Pope Gregory **XIII** in the 18th century. The Pope had also ordered that the leap year cycle was to be modified, ensuring that the leap year came 97 times in 400 years. The calendar which we currently follow is called the Gregorian calendar, named after Pope Gregory XIII.

Modern mathematicians found minute errors in the Gregorian calendar. It adds an extra 26 seconds each year. This would add up to a full day in a 3,320 year cycle.

Mathematicians and Astronomers say that even the Earth's speed around the Sun's orbit is not constant. Therefore, modern clocks today add leap seconds to make up for this discrepancy.

Scientists use trigonometry (a field of mathematics) to calculate the measure of the Earth's tilted axis and the effect it has on the lengths of days and nights, the timing and lengths of seasons.

Detailed graphs are plotted to understand the time of sunset and sunrise every day to predict tides, eclipses, etc. These calculations are also necessary to understand what day of the week a particular date will fall on in the future.

155 FAST FACT

MOST CALENDARS followed over the years are lunar based calendars because it is much easier to observe and follow the moon as compared to the Sun. This type of calendar is used traditionally to determine dates of religious events in countries such as India, China, and some parts of the Middle East.

156 NEW TYPES OF CALENDARS

🎓 **IN KEEPING WITH** modern global accounting practices mathematicians and economists devised new kinds of calendars to ensure the smooth continuity of business cycles and ease in book keeping methods. These calendars follow the Gregorian calendar but use different months as the beginning and end of the business year according to the norms of different countries.

THE 4-4-5 CALENDAR:
The first type of calendar is the 4-4-5 calendar. In this calendar, the year is divided into quarters. Each year consists of four quarters. Each quarter is made up of 13 weeks. This type of a calendar is very useful while making reports and financial statements, and enables efficient comparison of data over previous years. This calendar thus becomes a useful forecast for the management of business houses.

FINANCIAL CALENDAR/ 52-53 WEEK FISCAL CALENDAR:
The 52-53 week calendar is very useful for companies that want their fiscal or financial year to end on the same day every year. Companies prefer this calendar over standard Gregorian calendars for internal operations in order to break down their functioning into structured blocks of days/weeks, which remain consistent over several years. These companies follow the Gregorian calendar for all other external communication and administration purposes.

157 THE ANTIKYTHERA MECHANISM

ALMOST A CENTURY AGO, divers found an amazing mechanism on the sea bed close to the island of Antikythera. Research over the last few decades has revealed that it was a complicated devise used to study the cycles of the solar system.

Scientists agree that the Antikythera mechanism, which dates back to the 2nd century B.C., is considered to be the world's "First Computer."

The main contents of the wreckage included a wooden box with 30 bronze gears, which were connected to each other by minute teeth. Scientists have several theories about the functions of this mechanism. Most of them agree that it was an ingenious machine used to study planetary movements and predict the occurrence of natural phenomena like eclipses, etc.

The Antikythera mechanism has several inscriptions and pointers on its fragments, which continue to be studied by scientists. There are many dials that exist on the mechanism, which are considered to be calendars of different sets of years. For example, there is a 19 year calendar on one face, a 76 year calendar on another face, etc.

The discovery of the wreckage began an international project called the "Antikythera Mechanism Research Project" which brings together many scientists, government bodies, and museums.

158 CALENDAR FOR MARS

🎓 **MATHEMATICIANS AND ASTRONOMERS** have designed a calendar, which will enable them to keep a record of time on Mars. The principle of this calendar is the same as the Gregorian calendar used on Earth. This calendar has more months as Mars's orbit around the Sun is larger than the Earth's orbit.

The most important reason for developing this calendar was to enable planning for future human settlers on Mars. It is also an important tool for all the missions being sent to the red planet to study its environment.

Mathematicians calculated that an average Martian year would have 686.98 Earth days. They have also accounted for Martian leap years because the planet, like Earth follows an eccentric orbit. There were several challenges in making the Martian calendar, like tracking celestial events, determining the lunar period, measuring Martian seasons, determining the signs of the zodiac, etc.

Scientists have also developed formulas to convert Martian dates into Earth dates and vice versa for easy calendar calculations. Work is still being done to overcome these difficulties and come up with an error free calendar for Mars.

159 FAST FACT

📖 **THE EARLIEST KNOWN** date was 4236 B.C., which was found in the earliest Egyptian calendar.

160 FAST FACT

📖 **IN 1752,** the month of September had only 19 days in order to make up for the error of the Julian calendar. There were widespread riots by the general public protesting against this move, demanding that they be returned their 11 cancelled days.

161 FAST FACT

📖 **THE NAMES OF DAYS** of the week have been taken from seven celestial bodies. Sun (Sunday), Moon (Monday), Mars (Martis or Twesdaeg "Tuesday" in old English), Mercury (Mercurii or Wodnesdaeg "Wednesday" in old English), Jupitar (Jovis or Thunresdaeg "Thursday" in old English), Venus (Veneris or Frigedaeg "Friday" in old English), and Saturn (Saturday).

162 FAST FACT

📖 **IN THE HINDU CALENDAR,** the "Kalpa" is a measure of time that equals to 4,320,000,000 years!

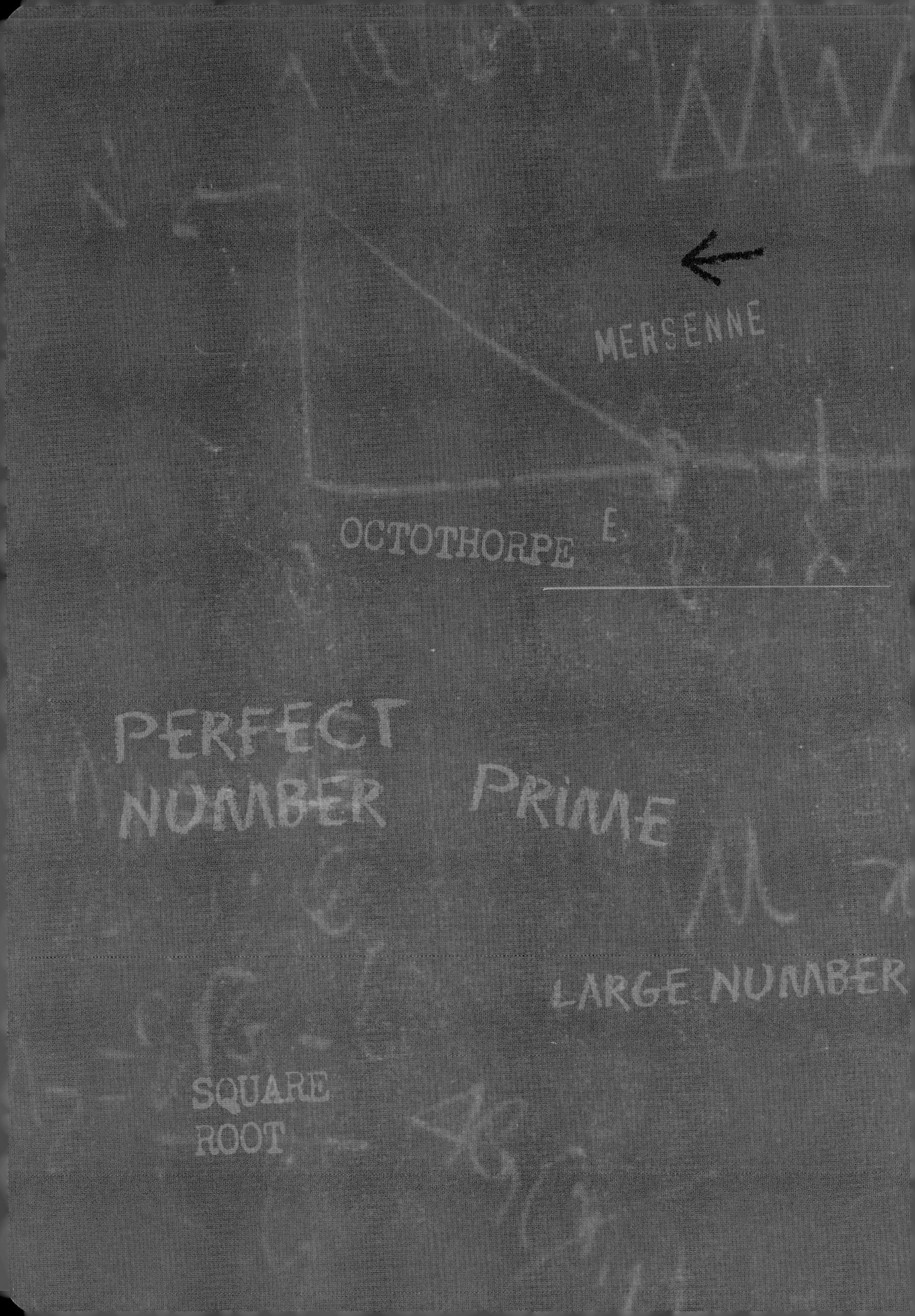

MERSENNE

OCTOTHORPE

PERFECT
NUMBER   PRIME

LARGE NUMBER

SQUARE
ROOT

501
Special Numbers and Shapes

163 ABOUT THE NUMBER E

THE NUMBER "e" is almost as important as "pi" π. The number was discovered quite recently, unlike π, which has been around since the Babylonian era.

e assumes great importance whenever growth and negative or positive has to be measured. One cannot work without e wherever economic or population growth is to be measured.

John Napier discovered a number that was constant. It constantly appeared in his work. However, the importance of this number became better known only in the 17th century through the works of Jacob Bernoulli on compound interest.

Mathematically, it can be proven that one unit of money will grow to e units in a year, if interest is compounded continuously. This number is sometimes known as Euler's number, because he denoted it as e while working with logarithms.

e is a transcendental number, i.e., it cannot be the answer to any algebraic equation. e to the power of pi is transcendental. Pi to the power of e being transcendental, is still a conjecture.

Around 1748, Euler calculated e to 23 digits. Euler further proved the earth-shattering identity linking real and imaginary numbers, e to the power of $i\pi + 1 = 0$. The number e was calculated to 10 to the power of 11 digits in 2007.

164 FAST FACT

THE NUMBER e is irrational (it is not a ratio of integers) and transcendental (it is not a root of a nonzero polynomial).

165 FAST FACT...

PI IS also a transcendental number. This could be proved only after the number e was proven transcendental. However, pi was known to mathematicians centuries before e.

166 FAST FACT...

SOME OF the most important numbers are approximations. These include root 2 (approximately 1.41421), pi (approximately 3.14159), phi (approximately 1.618) and e (approximately 2.71828).

167 MATHEMATICAL CONSTANT

MATHEMATICAL CONSTANTS ARE numbers which have a special significance. Euler's number "e" is a special number. Mathematical constants are found in many fields of math, like algebra, geometry, etc., and have an intrinsic mathematical importance in their fields.

Euler's number e is known as the exponential growth constant. The value of e exists in many mathematical formulas, such as those describing a nonlinear increase or decrease. Growth and decay, the statistical "bell curve," the shape of a hanging cable and a standing arch, all involve e. e also shows up in some problems of probability, some counting problems, and even the study of the distribution of prime numbers.

In the field of nondestructive evaluation, it is found in formulas such as those used to describe ultrasound attenuation in a material. The sound energy decays as it moves away from the sound source by a factor that is relative to e. Since it occurs naturally with some frequencies in the world, "e" is used as the base of natural logarithms.

168 FAST FACT...

EULER WAS also the first to use the letter e in 1727 (the fact that it is the first letter of his surname is coincidental). As a result, sometimes e is called the Euler Number, the Eulerian Number, or Napier's Constant. Euler proved that e is an irrational number, so its decimal expansion could not be terminated or be periodic.

An effective way to calculate the value of e is not to use the defining equation above, but to use the following infinite sum of factorials. Factorials are just products of numbers indicated by an exclamation mark. For instance, "four factorial" is written as "4!" and means $1 \times 2 \times 3 \times 4 = 24$.
$e = 1/0! + 1/1! + 1/2! + 1/3! + 1/4! + \ldots$

169 OF SQUARES, SQUARE ROOTS AND TRIANGLES

📖 **THE CONCEPT OF SQUARE ROOTS** was known to the Babylonians who compiled tables showing the square roots of numbers. Pythagoreans, in 525 B.C., arranged stones in the shapes of squares and triangles and studied their properties.

The squares so made are called perfect squares. The numbers of stones utilized to create these squares were all (1, 4, 16, 5, 36...) square numbers, while the numbers of stones used to make the triangles were triangular numbers. The sum of two successive triangular numbers was always a square number.

Since the square root of two is not a rational number, it immensely confused the Pythagoreans. However, two being the first even number, had its own special significance. Pythagoreans were aware that the sum of consecutive odd numbers is a perfect square.

A method for computing square roots was devised by Brahmagupta in 630 A.D. Yet, the symbol to denote a square root was introduced only in 1550.

If x^2 is plotted, the shape obtained is a parabola. A parabola is the shape of the curved path followed by objects falling to the Earth.

170 FAST FACT

📖 **RON GORDON** has named days in which the day and month are the square root of the year, for example 03.03.09. Gordon wants to promote the celebration of these days.

171 FAST FACT

📖 **IN ORDER** to blow up or reduce the size of an image without the use of computers, squares are drawn onto it and then each square is reproduced proportionately and according to the dimension required. This process is called graticulation. The process was also used by ancient Romans.

Squares always end in the numbers 0, 1, 4, 5, 6 or 9. The square of 1 is 1; while that of 11 is 121; of 111 is 12,321, of 1,111 is 1,23,432. This is true till the square of 11, 11, 11, 111 which is 12,34,56,78,98,76,54,321.

The square of 264 is 69,696. It is the only known undulating square, which means that its digits follow the pattern "ababab."

172 FAST FACT...

📖 **THE FECES** of wombats are cubic in shape.

173 THE AMAZING PI

🎓 **PI** (π) is the ratio of the circumference of a circle to its diameter. It is always constant. The number Pi has been around since Babylonian times. In 2000 B.C., the Babylonians were aware that its value was approximately three.

Several mathematicians have been working to find its accurate value to more decimal places. Aryabhata calculated pi as the fraction 62832/20000, which works out to be 3.1416. This correct to 4 decimal places. He knew that this was only an approximate value.

Archimedes said that the value of pi lies between 223/71 and 220/70, thus pointing us to 22/7, which is a value good enough for simple calculations.

Chinese mathematician Liu Hui found the accurate value to five decimal places. In the 5th century, Chinese mathematicians found its value to 10 decimal places.

Indian mathematician Ramanujam found its value to more than 30 decimal places. Johann Dayes, at the age of 20, worked on the problem mentally for two months and arrived on an answer that was correct up to 205 decimal places.

William Shanks worked on the problem for 20 years. In 1853, he finally claimed to find the correct value up to 607 places.

In 1949, the introduction of the computer proved that Shanks" value was correct only upto 527 places. The computers of that time calculated the value accurately to 2037 decimal places. By 2002, the value of pi had been calculated upto12411x10 to the power of 7 decimal places. This number is so large that if it were written across the equator, it would take 62 circles around the globe to write it.

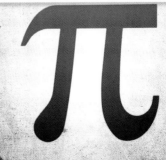

174 FAST FACT

📖 **PURE MATHEMATICIANS** spend time determining whether a series of numbers like 0, 1, 2, 3, 4, 5, 6, 7, 8, 9; or 10 6s or 10 7s fall together in the expansion of pi. Some mathematicians have even formed a society called "the Friends of Pi".

175 FAST FACT

📖 **MICHAEL KEITH** has written a remarkable poem modeled on the "Raven" by Edgar Allan Poe. The number of letters in each successive word of the poem gives us the value of pi up to 740 decimal places.

176 FAST FACT...

📖 **THE INDIANA** State Legislature once attempted to pass a bill to fix the value of pi.

177 FAST FACT...

📖 **LARRY SHAW**, a physicist at the San Francisco Exploratorium, started the celebration of Pi day on 14th March, 1988.

178 PERFECT NUMBERS

📖 **PYTHAGOREANS BELIEVED A NUMBER** to be perfect if the sum of the primes and the sum of the non-primes between one and the number was equal to the number. By this definition, 10 is a perfect number.

Pythagóras

Euclid, and later Nicomachus, studied perfect numbers in terms of relationships between the numbers and their divisors. They defined a number to be "superabundant" if the sum of its divisors other than itself is greater than the number. If it is less, it is said to be "deficient". According to them, every prime number is deficient, as the sum of the divisors of any prime number is always one.

A number is said to be perfect if the sum of the divisors of the number is equal to the number itself. Six is the first perfect number.

The next perfect number is 28. There are no perfect numbers between 28 and 496. Amazingly, it was found that every perfect number ends in six or 28.

Mersenne and Descartes were very interested in finding perfect numbers. Numbers in the form 2^{n-1} are called "Mersenne numbers." They are prime numbers that are instrumental in constructing perfect numbers.

Known perfect numbers are even. If an odd perfect number were to exist, it would have at least eight prime divisors, one of which would be greater than a million in value. This number would be at least 300 digits long.

179 FAST FACT

📖 **THE 48TH** perfect number was discovered on 25th January, 2013.

180 FAST FACT...

📖 **ST. AUGUSTINE** believed that the world was created in six days, as six is a perfect number.

181 FAST FACT...

📖 **THE 23rd** Mersenne number was celebrated by the release of a postage stamp by the University of Illinois, where it was found.

182 PRIME NUMBERS

PRIME NUMBERS are like the atoms of mathematics, from which mathematical compounds are made. Though prime numbers were known to Pythagoreans, they were discussed by mathematicians only about 200 years after the death of Pythagoras.

In 300 B.C., Euclid proved that there are an infinite number of prime numbers. Euclid also attested that every non-prime whole number greater than one can be written as the product of prime numbers.

Furthermore, there is only one way in which it can be done. This is known as "the fundamental theorem of arithmetic".

Euclid

183 FAST FACT...

TWO TO THE POWER of 756,839 – 1 is a prime number consisting of 227,832 digits found by the Cray-2 super computer.

However, there seems to be no pattern in the occurrence of prime numbers in the real number list. Eratosthenes worked out a method of sieving out prime numbers from a chart of counting numbers by crossing out numbers and their multiples, starting with two.

Whether a large number is a prime number or not can be difficult to say. Carl Friedrich Gauss, at the age of 15, suggested a formula to find the number of prime numbers existing before a given number. This is called "the prime number theorem", although the results received are known to be erroneous.

Christian Goldbach said that every even number greater than two is the sum of two prime numbers. However, there is no conclusive proof of this yet. The largest known prime number is called the "Mersenne prime", and is equal to 7.236x10 to the power 12.

Carl Gauss

Prime numbers can, however, be written as the product of complex numbers. For example, 5= (1-2i) (1+2i).

In the 18th century, it was discovered that 31,331,3,333,33,331,3,33,331,33,33,331, and 3,33,33,331 are all prime numbers. However, 33,33,33,331 is not a prime number.

184 ABOUT LARGE NUMBERS

🎓 **AN INFINITE NUMBER OF RATIONAL** numbers (fractions) and irrational numbers exist between any two closest points on the real number line. The Mersenne prime is a prime number whose value is $(7.236 \times 10)^{12}$, which is approximately 7 million.

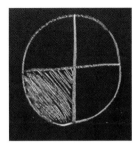

The Fermat prime (F1945) represents a number so huge that it is larger than the sum of all the particles in the universe. In 1995, a patent was claimed by an American mathematician on two very large prime numbers for use in securing computer systems. American and British names for large numbers differ.

The Americans give names to numbers based on powers of 10 that are multiples of three; while the British use multiples of six. Thus, in the American system, 10 to the power of six is a million, 10 to the power of nine is a billion, 10 to the power of 12 is a trillion, 10 to the power of 15 is a quadrillion, and 10 to the power of 18 is a quintillion.

On the other hand, the British call 10 to the power of 12 a billion and 10 to the power of 18 a trillion. So, a British billion is 1000 times larger than the American one. Thus, inter-country trade can get a little confusing sometimes.

Myth has it that a courtier in India named Sissa was asked to chose a reward by the king. Sissa had recently invented a game similar to chess. As a reward, he asked for grains of rice. The condition was that he wanted one grain of rice for the first square on a board, and the double of the previous number of grains for each successive square.

Sissa's request seemed to be an extremely small quantity to all. The king thought that Sissa was mad and happily ordered his request to be fulfilled. However, to the amazement of all, the numbers grew so large that it was impossible to fulfill Sissa's condition.

185 MAGIC NUMBERS

MATHEMATICIANS CONSIDER SOME NUMBERS to be "magic numbers". Numbers like 2, 8, 20, 28, 50, 82, and 126 are considered magic numbers, because nuclei having these many protons or neutrons are stable.

The number 1,729 was called the magic number by mathematicians G.H. Hardy and Srinivasa Ramanujam, because it is the only number which can be expressed as the sum of the cubes of two different sets of numbers. 10 to the power of 3 + 10 to the power of 3 sums up to 1729, and 12 to the power of 3 and 1 to the power of 3 sums up to the same number.

186 FAST FACT

WORDS REPRESENTING large numbers like "million" were not a part of the original English number system. They are actually the result of Latin influence on the language.

187 FAST FACT

VINCULUM is the name of the division bar. Virgule is the division slash. Octothorpe is the number sign. Obelus is the division sign. Residues of ancient counting by 20s and 12s are the words score, dozen, and gross.

188 FAST FACT...

THE NUMBER of letters in the successive words of the sentence "May I have a large container of coffee?" corresponds to the value of Pi (3.1415926).

189 GOOGOL - AN EXTREMELY LARGE NUMBER

GOOGOL is the name given to the number formed by a one followed by a 100 zeros. This name was suggested by Milton Sirotta, the 9 year-old nephew of mathematician Kasner.

While on a walk with his nephew, Kasner asked him for a name for the extremely large number. The child, Sirotta, understood that since the number was finite even though enormous, it had to have a name, and suggested the name "Googol." The term was first used in the book "The Imagination," in 1940.

Milton Sirotta exhibited, along with the innocence and intuitiveness of a child, an immense understanding of the logic of numbers by suggesting the name googol for a number that was made up of a one followed by writing zeros till the writer got tired. Of course, this number would be finite, as any person would get tired eventually.

However, there was a problem in assigning a specific value to this number, since different people would get tired after writing different numbers of zeros.

Kasner gave it finiteness by defining a googolplex to be a googol followed by a googol of zeroes. Thus a googolplex is 10 raised to the power of googol.

190 FAST FACT

THE FAMOUS search-engine Google owes its name to an incorrect spelling of googol. The name was chosen to convey that the search-engine would throw up a large amount of data. The name of the headquarters for Google (the company) in California is Googleplex.

191 FAST FACT...

THE NUMBER googol was the subject of the question that was to fetch Charles Ingram a million dollars in the show "Who wants to be a millionaire?" in 2001. He answered the question correctly, but did not receive the million dollars as he had cheated by taking help from his wife and another contestant.

192 FAST FACT...

THE NUMBER googol is so large that if a zero was put on every inch of the way to the farthest star, it would still be impossible to put down all the zeroes in the number.

193 NUMBERS HAVE FRIENDS TOO

SOME PAIRS OF NUMBERS ARE "amicable," which means that the proper divisors (a proper divisor of a number is a positive factor of that number other than the number itself) of one adds up to the proper divisors of the other. For example, the proper divisors of 220 are 1, 2, 4, 5, 10, 11, 20, 22, 44, 55, and 110. The sum of these numbers is 284.

On the other hand, the proper divisors of 284 are 1, 2, 4, 71, and 142, and the sum of these is 220. This is the smallest pair of amicable numbers.

Other pairs are 1184 and 1210, 2620 and 2924, 5020 and 5564, 6232 and 6368, 17296 and 18416 and 9363584 and 9437056, etc. With the formulation of equations to find amicable numbers and the advent of computers, many pairs of amicable numbers have been discovered. In 2007, there were almost 12,000,000 known amicable pairs.

194 FAST FACT...

IN EVERY known case, the numbers of a pair of amicable numbers are either both even or both odd.

195 THE GOLDEN RATIO

MATHEMATICIANS HAVE been enamored with what has been known by several names, like the golden ratio, the golden number, the golden section, the golden proportion, the golden mean, and the divine proportion.

The earliest written record of the golden ratio is found in a definition by Euclid around the 3rd century B.C., where he calls it the extreme and mean ratio. According to him, two quantities are said to be in a golden ratio if the ratio of the sum of those quantities to the bigger of them is the same as the ratio of the bigger quantity to the smaller. That is to say, if "a" and "b" are two quantities such that "a" is greater than "b", they are in a golden ratio if "a+b": a =a : b. This ratio was worked out to be approximately 1.618.

Johannes Kepler

Things that are in a golden ratio are aesthetically pleasing to the human eye. Knowingly or un-knowingly, architects, artists, and sculptors have since ancient times incorporated the ratio in their works.

The sculptures created by Phidias, which date back to the 4th century B.C., are in this ratio. The pillars and beams of the Parthenon in Greece are also shaped according to this ratio. The ratio is also seen in the pyramids of Egypt.

Johannes Kepler, a famous mathematician, called the ratio "a valuable jewel" and the Pythagoras theorem as valuable as "gold". He combined the two into what is now known as Kepler's triangle. In the 20th century, the symbol phi (φ), which is used for the golden ratio, was coined by Mark Barr to honor Phidias.

196 FAST FACT

ULTRA SLEEK machines like the spaceship USS Enterprise "Ain" the movie Star Trek and the Aston Martin driven by James Bond have incorporated this ratio in their designing.

197 FAST FACT

OUR CREDIT cards are in the shape of rectangles whose sides are in a golden ratio. Such rectangles are called golden rectangles.

198 FAST FACT

IF A SQUARE is cut away from a golden rectangle, the resultant rectangle is once again a golden rectangle. This is true, no matter how many times the process is repeated.

199 THE FIBONACCI SEQUENCE

🎓 **THE FIBONACCI SEQUENCE** is the sequence of numbers 0, 1, 1, 2, 3, 5, 8, 13, 21, 34, 55, 89. In the "Fibonacci sequence", each successive number is the sum of the previous two numbers.

It is strange that Fibonacci arrived at this sequence by pondering over the breeding of rabbits. Fibonacci began with one pair of young hypothetical rabbits. He set out unnatural conditions that all rabbits pair once a month and produce a pair of offspring – a female and a male, at the end of every month, and none die.

Therefore, at the end of the 1st month, there would be one mature rabbit pair who would reproduce. That makes two pairs, one mature and one young. The mature would reproduce, so there would be three rabbit pairs, and so on. He wondered how many rabbits there would be at the end of a year.

Work on this word problem led Fibonacci to formulate the sequence, which made him famous. Fibonacci set out the rabbit breeding problem in his book. While he was also known as Leonardo of Pisa, he was better known as Fibonacci, which is the name used to immortalize the sequence.

200 FAST FACT

📖 **WHAT IS** common between the ratio written out in 300 B.C. and the Fibonacci sequence? In the 15th century, Johannes Kepler proved that, the ratio of any two successive terms of the Fibonacci series is almost the same as the golden ratio.

201 FAST FACT

📖 **THE HIGHER** one goes in the Fibonacci series, the closer he she gets to the golden ratio.

202. NATURE AND THE FIBONACCI SEQUENCE

THE FIBONACCI SEQUENCE is amazing because it turns up in diverse places. It is found in nature, in the number of spirals in sunflowers (34 in one direction and 55 in another), in the curves of the arms of galaxies, and in the organization of DNA molecules.

The sequence is also found in the proportions of famous sculptures like Phidias and Peter Randall, whose sculptures Page and Seed are based on this sequence. The sequence is also seen in musical compositions. The "Dance Suite" by Bartok is inspired by the Fibonacci sequence.

The sex of a baby bee depends upon whether it was born from a fertilized or unfertilized egg. If the egg was unfertilized, the baby is a male or drone bee. If the egg is fertilized, then the baby is female. Thus, a male bee will have only one parent. However, it will have two grandparents, three great-grandparents, five great-great- grandparents, and so on. Extending this sequence backwards, the ancestors of a male bee make up the Fibonacci sequence.

Mathematicians find it interesting to discover various relations in the successive numbers of the Fibonacci series. They have found that if a number of terms of the series are added, the sum is the same as one less than the number after the next number in the series.

The sum of any 10 numbers of the series is equal to 11 times the seventh of the numbers chosen. They also found that if coins of the denomination of one and two are placed in a bag, the number of ways coins can total to four in value, is five. This is the fifth number in the famous series.

203. FAST FACT...

THE MODERN world attributes this sequence to Fibonacci. He probably picked up this idea from India. Indian mathematicians like Gopala, Pingala, and Hemachandra used the series centuries before Fibonacci wrote about it. It was only as late as the 19th century that Édouard Lucas named it the "Fibonacci sequence".

204. FAST FACT...

IN THE BEST-SELLING book "the Da Vinci Code", the murder victim wrote a part of the Fibonacci sequence in his blood as a clue for the investigators.

205 THE DIVINE PROPORTION

🎓 **LUCA PACIOLI** was a monk, an accountant, and the teacher of Leonardo da Vinci. He could see similarities in the characteristics of God and the golden ratio. Pacioli said, "Like God, the proportion is always similar to itself." In 1509, he called it the Divine proportion.

Influenced by his teacher and the ideas of Vitruvius Pollio about the geometrical proportions of the ideal man, Da Vinci incorporated the ratio in his drawings. In his famous drawing, the "Vitruvian Man", two superimposed human male figures are drawn simultaneously within a circle and a square.

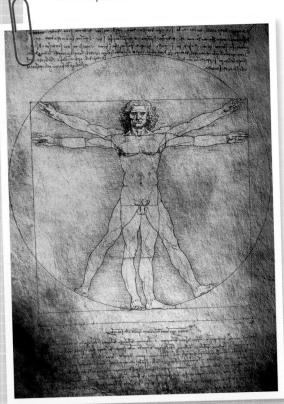

The radius of the circle is in a golden ratio to the side of the square. The body parts of the Vitruvian Man are also in a golden ratio/Fibonacci series. This is demonstrated by the lines drawn across the figure. The drawing of the Vitruvian Man inside a circle and a square at the same time is seen as an attempt to solve the mathematical problem of "squaring a circle" (constructing a square with an area equal to that of a given circle using a compass and a ruler only). This problem occupied the minds of mathematicians of the time.

Leonardo's painting was named the Vitruvian Man to honor Vitruvius, whose ideas about proportionality accompany the drawing.

206 SPIRAL: THE MAGICAL SHAPE

IN MATHEMATICS, a spiral is a curve that emanates from a central point, getting progressively farther away from the point as it revolves around it. The spiral shape is repetitively seen in nature, from a micro to a macro level.

The shape also appears in such diverse aspects of nature such as DNA, shells of snails, paths of hurricanes, and paths of hawks as they approach their prey, and the branching of human nerves. What is interesting to know is that this spiral can be drawn using the Fibonacci sequence.

The first two adjacent squares with sides equal to unit one are drawn, the first two numbers in the sequence being one. Then, a square with side equal to unit two (the next number in the sequence) is drawn with one side along the longer side of the rectangle formed in the previous step. The third square is drawn using the third number in the sequence, which is three, again with one side along the longer side of the rectangle formed in the previous step. The next square is of unit five, and similarly drawn. This process may be repeated indefinitely. The curve joining the opposite corners of the successive squares forms the golden spiral.

A spiral in which the point tracing the spiral gets further away from the originating point by the golden ratio is called the golden spiral.

If a curve is drawn connecting the opposite corners of the golden rectangles is drawn, we get the golden spiral once again.

207 FAST FACT

OUR OWN GALAXY, the Milky Way, is a barred spiral with several spiral arms.

208 FAST FACT

THE CHAKRAVYUHA, a military formation mentioned in the Indian epic "Mahabharata" is a rotating spiral formed by soldiers.

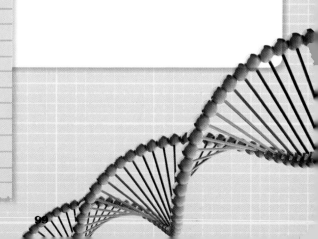

209 THE RATIO IN ART

🎓 **APART FROM** the golden ratio, other ratios too have been used by artists and architects. Ancient artists, as a rule of thumb, depicted the height of an average person (including the head) as seven-and-a-half times the person's head.

Nelson Mandela

Gods and superheroes were shown to be eight-and-a-half heads tall by giving them longer torsos. They also matched nature to make the human foot almost exactly the same length as the arm from wrist to elbow. Human eyes were always situated in the middle of the face.

A wonderful example of the marriage of art and math is a sculpture of Nelson Mandela's face made using 50 steel bars. Sculptor Marco Cianfanelli used steel bars of varying thickness, reminiscent of barcodes, to construct the same. When viewed from the correct angle of exactly 90^0, the sculpture shows Mandela's face behind bars, to depict Mandela's 27 years spent in prison. This sculpture is made of 50 rods, each symbolizing a year since the arrest of the leader fifty years ago.

210 FAST FACT...

📖 **IF WE PLOT** the curve represented by the equation $(x^2+y^2-1)^3-x^2y^3=0$, the resultant curve would be in the shape of a heart.

211 GREEK MATHEMATICIANS AND INFINITY

🎓 **THE WORD** "apeiron" was used in ancient Greece to describe the concept of no boundaries. It is believed that Greek mathematicians were the first ones to accept the concept of infinity after observing three traditional unchanging concepts – time was endless, space and time could be divided without end, and that space was seemingly boundless.

An understanding of the fact that a circle can have unlimited polygons inscribed in it made it necessary for the Greeks to ponder over the concept of infinity. Theorems which state that the number of primes was limitless led the Greeks to understand that the total numbers also needed to be limitless.

The Greeks, however were unable to clearly define infinity, which led to a limited development in their math skills. Many mathematicians created paradoxes by combining finite and infinite processes. But neither the Greeks nor any other culture seemed to be concerned about the fact that time did not have a defined beginning or end!

The Greeks, however, were not inclined to use any form of incommensurables to any extent. The quest to understand the concept of infinity continued over the coming generations, leading to a great progress in the field of mathematics.

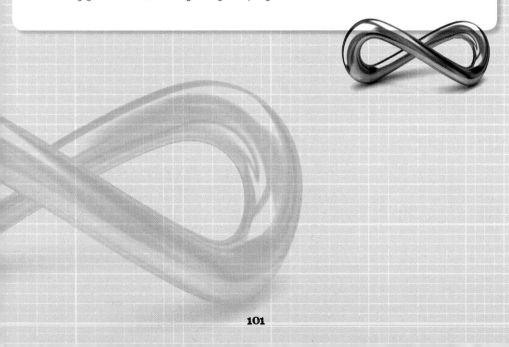

212 ARISTOTLE AND THE CONCEPT OF INFINITY

🎓 **ARISTOTLE WAS A GREEK PHILOSOPHER** born in the year 384 B.C. He studied in Plato's academy and later became a tutor of Alexander the Great.

Aristotle was an astronomer and a scientist who had a deep liking for the science of meteorology. He also examined human psychology. Aristotle's work on how humans perceive the world continue to underline many principles of psychology even today.

Aristotle maintained a cautious view on the concept of infinity by defining "minimal infinity". The term enabled solving equations and proving theorems, and ensured that no controversies arose by defining a new number. This potential helped mathematicians for the next couple of millennia.

Aristotle believed that many magnitudes of numbers cannot be infinite, because adding such large quantities would then cross the limits of the universe! His thinking, like other Greeks of his time, was that infinity is imperfect and unthinkable.

Even in geometry, Aristotle was of the opinion that points are on a line segment but do not comprise the line meaning that the "continuous cannot be made of the discreet."

Aristotle

213 FAST FACT

📖 **ARISTOTLE WOULD** walk in the school grounds while teaching children, who were forced to walk briskly to keep up with him. They were nicknamed "Peripatetics" meaning "people who travel a lot" for this habit.

214 TIMELINE – THE CONCEPT OF INFINITY

🎓 **IN THE 1400s**, the Europeans adopted the concept of God being infinite.

In the 1500s, Galileo mentioned the word infinity in his statement "It is wrong to speak of infinite quantities as being the one greater or less than or equal to the other." Mathematicians have interpreted this as Galileo's acceptance of infinity obeying many rules.

In 1544, Fibonacci wrote the book, "Arithmetica Integra" and mentioned that irrational numbers were not true, and hid in the cloud of infinity.

Galileo

In the 1500s, Steven Simon's work on the division of a triangle by median accepted the concept of limited infinity.

In 1657, John Wallis introduced the symbol of infinity that we use even today which denotes an unending curve.

In the 1700s, Newton, Leibniz, Bernoulli, George Berkeley, and Euler debated about the concept with limited results.

In the 1800s, several mathematicians like Bolzano, Cauchy, Dirichlet, Gauss, Weierstrass, Riemann, and Dedekind studied the concept of infinity and gave their viewpoints.

In 1874, the concept of infinity was brought into clearer light from the work done by Georg Cantor.

215 GEORG CANTOR – INFINITY AND SET THEORY

GEORG CANTOR was a German mathematician famous for his development of the "set theory". In mathematics, the set theory is considered to be one of the most important concepts. His work clearly defined the "one to one" correspondence between the elements of two sets, infinite sets, and well-ordered sets. His work on cardinal and ordinal numbers was well-accepted by all mathematicians, though certain other concepts did get immense opposition.

Cantor proved that every set has a power set (a set of all its subsets) that is strictly bigger than the given set. This meant that the power set could not be put in a one to one correspondence with the given set, even in the case of an infinite set. This proof was easiest to understand in terms of real numbers.

Some mathematicians however, rejected this "Cantor's Paradise", calling it ill-defined.

Cantor was recognized and awarded the highest of recognitions from mathematical societies for his phenomenal work and contributions to the field of mathematics. Renowned mathematician David Hilbert fiercely defended Cantor whenever he was criticized for his work on transfinite numbers.

216 FAST FACT

THE SYMBOL of infinity is interpreted by some as a snake biting its own tail!

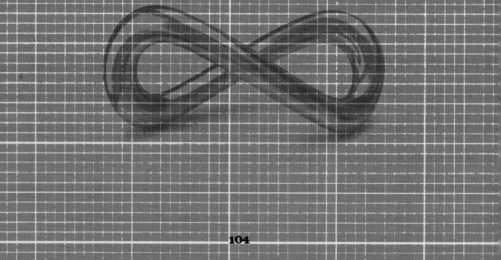

217 TRIANGLES

🎓 **IN 1070**, triangles were discovered by Omar Khayyam, who is better known for his poetry. In some countries, the triangle is named after him.

The theorem is attributed to Pythagoras, but known for centuries before him is a famous theorem relating to triangles. There are several proofs of this theorem, but the most used one is the one by Euclid.

The one given by Bhaskara in the 12th century is a very famous one. According to him, triangles of equal areas are cut out from a square of sides a and b, and it is demonstrated that $a^2+b^2=c^2$.

A triangle has several centers, such as the centroid, the orthocenter, and the circumcenter. All of these coincide in an equilateral triangle.

Napoleon Bonaparte's skill at mathematics got him into artillery school, and he is known to have enjoyed the company of mathematicians while he was the emperor.

Little evidence regarding his skill as a mathematician is known. However, Napoleon has the honor of having a theorem attributed to him, published a few years after his death in St. Helena.

Napoleon probably had little to do with its discovery or proof, but Napoleon's theorem concerns the construction of equilateral triangles.

Surveying of land is done in modern times by using triangles as opposed to squares. The great trigonometrical survey of India used this method. It began in 1800 and went on for 40 years!

The triangle has a more rigid shape than a square or rectangle and is therefore used in construction. The joining together of triangles creates a truss.

Designs based on equilateral triangles are more stable than those based on isosceles triangles. The Warren truss, using equilateral triangles, was used to construct the London Bridge Station in 1850.

218 FAST FACT

📖 **THE BERMUDA** triangle is a triangular portion of the sea between Miami, Bermuda, and Puerto Rico where ships and aircrafts have known to mysteriously disappear.

219 PASCAL'S TRIANGLE

🎓 **THE TRIANGLE** made by writing 11 and the values of the ascending powers of 11 in subsequent rows gives us the Pascal's triangle.

Blaise Pascal, in 1651, realized that there were several hidden relationships within this triangle. A single sheet of paper was not enough to discuss all of them. Pascal's paper on the properties of this triangle was published after his death.

Beginning with one in the first row and placing two number ones in the next row on either side of the first one is the beginning of Pascal's triangle. The next rows are found by writing the number one on either side of the row and filling in the gaps by writing the sum of the two numbers immediately above them.

220 FAST FACT

📖 **ALTHOUGH**, named after Pascal, this number triangle was known to the ancient Indians and ancient Chinese.

221 FAST FACT

📖 **LEIBNIZ'S** triangle is formulated by adding the numbers in the row below it. It also displays symmetry in the form of a harmonic series. It can also be related mathematically to Pascal's numbers.

The symmetries exhibited by this triangle are fascinating. It has mirror symmetry, i.e., the vertical line down its center divides it.

Counting numbers, triangular numbers, and tetrahedral numbers can be read off different diagonals of this triangle. There are partial diagonals that show symmetry.

The interlacing of such partial diagonals results in a Fibonacci sequence!

Pascal's triangle also gives one a number of ways in which certain objects may be combined from a larger number of objects. For example, there are many ways we can choose four objects out of nine different objects. C (n, r) is the number in the nth row at r position. These are called Pascal's numbers.

If all the odd numbers in a triangle are substituted by one and all even by zero, the result is a pattern that will be a fractal.

Pascal

222 TRIANGLES

🎓 **THE WORD "FRACTAL"** comes from the Latin word "fractus" meaning irregular or fragmented. Fractals are visual representations of mathematical formulas. They were discovered by accident in 1980 when a printing device printed dots as per the commands of a computer at the IBM research center.

The emerging fractal was named the "Mandelbrot set" after Benoit Mandelbrot, who first saw the beauty of the pattern. It was a curious case where the result, i.e., the image, was recognized before the theory that defined it.

Mandelbrot coined the word "fractal" but had no mathematical definition for it. Defining the fractal would have to be done carefully so as to not limit it in any way. He believed that the idea of fractals needed to mature like wine before it could be defined.

223 FAST FACT

📖 **THE NAUTILUS** shell is the most perfect fractal in nature. It is also a Fibonacci spiral.

The simple formula "$x^2 + c$" generates the Mandelbrot set. Fractals are generated by applying the same formula over and over again.

Zooming in on a Mandelbrot set will give the same Mandelbrot set. This was known before the fractal was recognized. While surveying a coastline, the pattern is repeated more often if a smaller unit is chosen. In 1879 and 1904, mathematicians had worked on "fractals" without the use of a computer. Their work was not carried to its conclusion.

A fractal that looks like a snowflake was discovered by Koch in 1904. It is the first known fractal curve. The curve is formed by splitting the sides of a triangle into three parts and adding a triangle in the center.

The area of the triangle remains the same, but the length increases. It can be said that the curve has a finite area but an infinite circumference. The fractal named "Sierpinski gasket" was found by subtracting triangles from an equilateral triangle. If a line, square or a cube is scaled up, its length, area, and volume increase. However, if the basic unit of a fractal is scaled up, it usually does not increase its dimensions, as was demonstrated in Koch's curve. In nature, fractals are seen in cloud formations, plants, corals, and sponges. Surprisingly, it is also seen in the spread of modern cities.

224 WHAT'S YOUR ANGLE

🎓 **ANGLES ARE FORMED** when two lines meet. They are used to measure the degree to which any line or object has turned. According to Greek astronomer Hipparchus, a full turn was to be divided into 360°. He studied the properties of the Earth as a sphere.

Hipparchus conceived the idea of imaginary lines drawn from pole to pole at known angles to each other (longitudes), and in parallel bands (which may not make any angle) perpendicular to longitudes. He did so in order to determine the position of places on the surface of this sphere.

The relationship between angles of triangles and that of circles was studied by ancient Greeks. Angles of elevation helped them understand the position of stars.

The Greeks even calculated the circumference of the earth quite accurately using angles cast by shadows at different places on the same latitude. Many of these rules are still used in scientific calculations.

Navigators position themselves using degrees. Compasses use angles to tell us the position of the magnetic North pole. Knowing this is imperative to navigate ships.

The entire discipline of trigonometry is based on the study of the angles and sides of a triangle. The angle made by a bubble inside a spirit level is used to determine if a surface is flat.

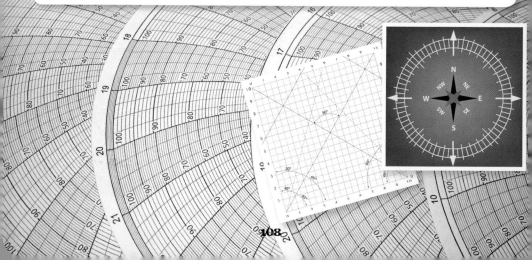

225 FAST FACT

📖 **IF A TRIANGLE** is drawn on a sphere with intrinsic straight lines, the sum of the angles of such a triangle will be greater than 180^0.

226 FAST FACT

📖 **PHOTOS TAKEN** from unusual angles make very interesting pictures.

227 FAST FACT

📖 **IN SPITE** of the extensive use of the metric system in all other units of measurement, angles are still measured in degrees and not grads (100 grads being equal to 90^0).

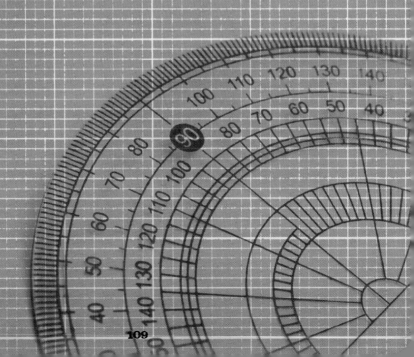

228 SHAPES THAT FIT PERFECTLY

🎓 **FLAT SHAPES** with three or more sides are polygons. Polygons whose sides and interior angles are equal are regular polygons. Regular polygons occur frequently in nature.

Henzene molecules are hexagons. The word "hexagon" comes from the two words "hexa" meaning six and "gonia" meaning angles.

The cells of honeycombs are hexagonal in shape as well. These hexagons fit together perfectly and thus prevent the entry of dirt and predators. Cells in the shape of triangles or squares would have fitted perfectly too. In fact, triangles provide great strength in constructions. However, hexagons provide the largest volume per unit of wax used. Therefore, economic considerations make the bees choose the hexagon.

Regular shapes that fit perfectly into each other are said to tessellate. We use this process to construct our bathrooms with tiles that fit into each other perfectly so that water cannot seep through. Structures of crystals have distinct tessellations and are used by scientists to identify them. Ice crystals are also arranged in hexagonal structures.

Tessellations can also be done using more than one shape. They can be done on non-plane surfaces as well. For example, pentagons and hexagons are used together to make three-dimensional tessellations on a soccer ball.

Artist M.C. Escher used tessellations and optical illusions to create what may be called 'mathematical art'. He has made a piece called "circle limit IV", in which figures of devils and angels are tessellated in a circle. Optical illusion makes the piece look like a sphere. Islamic artists have used tessellations to decorate the doorways and interiors of many buildings.

229 COLORING MAPS

IN MAPS, shapes of the countries, counties or districts may be irregular but they fit perfectly into each other. They share boundaries with various other countries, counties or districts.

Mathematician De Morgan presented his young neighbor, Tim, with a box of crayons and asked him to color a map of England in such a way that counties with a common border were colored using different colors. Tim thought that the number of colors in his box were too few to do so. This set his mathematical brains thinking.

The minimum number of colors required to do so was a thought provoking issue for mathematicians. It was believed that four colors were enough but mathematicians required proof.

It is believed that Sanders Pierce had found a solution in 1860, but unfortunately, if it existed it is no longer available. Mathematician Arthur Cayley tried but failed to find a solution. His student Alfred Bray Kempe published a solution soon after and was granted membership to the Royal Society of London.

230 FAST FACT
THE MATHEMATICS department of the University of Indiana honored the discovery of this proof by a postage stamp which proclaimed "four colors suffice".

However, it was found to be false 10 years later by Percy Heawood, who found a map to which the proof did not apply. Mathematicians could not find a map that could be colored with four colors, yet in the absence of proof that four colors were enough, mathematicians continued the search.

The problem was then expanded to include maps not drawn of a single plane or sphere. Maps drawn on three dimensional shapes with holes in the middle began to be considered.

Heawood proved a formula that gave the number of colors guaranteed to color maps with "n" number of holes. However, he did not prove this to be the minimum number of colors. In this formula, if the number of holes taken is one, the number of colors is found to be four!

In June 1976, with the help of IB computers, a proof of the four color problem was derived. However, since the proof was very long, it could not be checked by humans, and there still remains a possibility of error in the minds of some mathematicians.

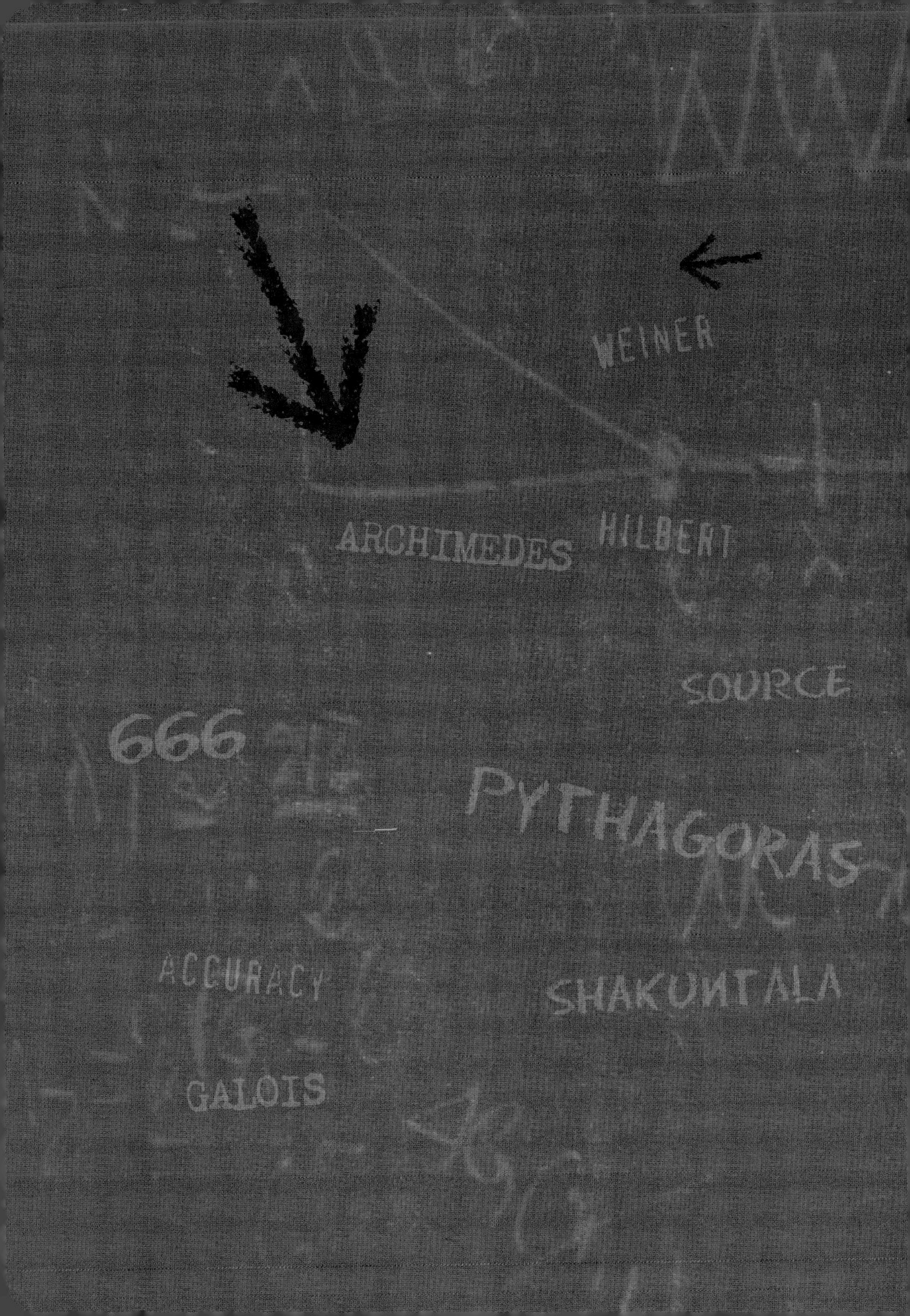

WEINER

ARCHIMEDES HILBERT

SOURCE

666

PYTHAGORAS

ACCURACY SHAKUNTALA

GALOIS

501 Mathematicians, Lives, Deaths, Eccentricities and Superstitions

231 CALCULATING AT THE SPEED OF LIGHT

A NUMBER OF PEOPLE cannot commit numbers larger than seven digits to memory. Several prodigies can, from a very young age, do mathematical calculations faster than comprehensible to normal adult minds.

The first documented rapid mathematician is known to have been Jedediah Buxton, a farm laborer from Derbyshire.

Jedediah was able to conclude after a performance of a play called "Richard III" that a total of 14,445 words had been spoken and that the actors had taken a total of 5,202 steps.

232 FAST FACT

THOMAS FULLER and Jedediah Buxton both did not know how to read and write.

Another prodigy, Thomas Fuller, was an African slave in Virginia, U.S.A. When he was 14 years old, he took only a minute and a half to calculate accurately the number of seconds a man who was 70 years, 17 days and 12 hours old would have lived. Thomas even remembered to add the additional seconds for leap years.

An observation was made that not all prodigies, including mathematical prodigies, have high IQs. It was also noticed that a large percentage of these prodigies exhibit autistic traits such as attention to detail, low social and communicative skills.

Several years ago, children with such brilliant minds were sometimes considered stupid. Modern technology like PET scan suggests that mathematical prodigies have exceptionally long-term working memories or LTWM.

233 FAST FACT...

SHAKUNTALA DEVI was a prodigy from India, popularly known as the "human computer". She calculated the 23rd root of a 201 digit number in the 1970s within a good 50 seconds. It took a computer a full minute to confirm her calculation.

Therefore, such children can not only remember large strings of numbers, but can also process a large amount of verbal and non-verbal information inside their heads.

234 PYTHAGORAS AND HIS BROTHERHOOD

BESIDES BEING A MATHEMATICIAN, Pythagoras was a mystic and a prophet. He had formed a cult-like group, "the Brotherhood of Pythagoreans", who believed that numbers rule the universe.

Πυθαγόρας
Pythagóras

The brotherhood had their own symbols, rituals and prayers, and was extremely secretive about their findings. As a result, it is not clear who from the brotherhood actually proved the Pythagoras theorem.

When Pythagoreans were unable to determine the exact value of root 2 which is the value of the hypotenuse of a right angled triangle of unit side, they were extremely upset.

The brotherhood was so shocked at the existence of a number which could not be represented as a fraction, that they called it and all such numbers "alogon" meaning "unutterable" and did not speak of their existence to the world. Ironically, root 2 is known by some as a Pythagorean number.

235 FAST FACT...

THE BROTHERHOOD of Pythagoreans believed in rebirth. They further believed that eating animal products was cannibalism, as they might contain the trans-migrated souls of human ancestors.

236 FAST FACT...

PYTHAGORAS IS BELIEVED to have been murdered by people who were displeased by the power he had gained.

The Pythagoreans believed numbers had qualities. According to them, the number one had spiritual existence. They also saw 10 as a perfect number because the number of prime and non-prime numbers between 1 and 10 is equal. A number, whose factors included one but not itself and added up to the number, was also seen as perfect numbers by the Pythagoreans.

Pythagoreans discovered that pleasing musical notes were produced when blacksmiths used hammers whose sizes were in proportion, thus establishing a connection between music and mathematics.

237 THE MATHEMATICS UNDERLYING: ALICE IN WONDERLAND

🎓 **ALICE IN WONDERLAND** was written by Lewis Carroll (pen name of mathematician Charles Lutwidge Dodgson) as a satire aimed at becoming abstract.

238 FAST FACT...

📖 **OTHER MATHEMATICIANS** were good writers too. Omar Khayyam is renowned for having written the "Rubaiyat". Very few people knew that he was a mathematician and wrote a book on Algebra.

Various mathematical concepts have been incorporated by Lewis Carroll in his book, Alice in Wonderland. The seemingly simple story is actually a satire on new math, which used new methods to prove that two and two did not add up to four.

Mathematical ideas form the basis of incidents that seem to be just nonsensical fun. For example, when Alice finds herself getting smaller and smaller, she wonders if she will reach the point of nothingness. This alludes to the idea of finiteness.

In the tea-party scene, Carroll talks about time being the fourth dimension. Alice is confused as multiplication tables slip out of the base 10 number system. Alice has to figure out how to stay gracefully geometric, having the same proportions at any size, before she can go on.

239 JOHN VON NEUMANN

THE OBSERVATION that the game of poker depends not just on probability but also on "bluffing" led John von Neumann to formulate his famous Minimax theorem of Game theory.

Game theory was used by Neumann to plan the path that should be taken by American bombers to minimize the chances of being detected.

John von Neumann was the son of a Jewish father. John was originally named János Neumann but was called Jancsi as a child. In 1913, his father purchased a title but did not add "von" to his name.

However, his son did to become János von Neumann. When he came to the United States, he began to be called Johnny. It is said that he was sure to come up with a solution of any problem left unsolved in class and would scribble it on a piece of paper.

John Neumann and his second wife Klara Dan had a very different life than what would be expected from a mathematician of his caliber. He hosted parties which went on late into the night. Even when he was in Germany, he had always enjoyed the night life.

240 FAST FACT

ZERAH COLBURN and Truman Safford were both child prodigies born in the 1800s at Vermont, but it is said that they both lost their calculating powers as they aged.

241 RAMANUJAN

🎓 **THE STORY OF MATHEMATICIAN** Srinivasa Ramanujan, a clerk from India and a college dropout, begins in 1913.

Ramanujan wrote a 10 page letter to mathematician G.H. Hardy. It was filled with about 120 statements about infinite series, improper fractions, and number theory. On going through the letter with his friend J.E. Littlewood, they were forced to conclude that the results derived in the letter must be true.

Ramanujan had always been enchanted by the field of mathematics. He needed monetary support to be able to carry on his mathematical studies. Mathematician Ramchandra Rao recognized Ramanujan's potential and provided him with a small subsidy. He even got him a job as a clerk, which allowed him to work on mathematics, scribbling theorems on scraps of paper. Ramanujan wrote an article about Bernoulli's numbers which was published in the Journal of Indian Mathematical Society. However, his genius was not recognized fully until Hardy got in touch with him.

Hardy wanted Ramanujan to come to England to continue his work. After a lot of persuasion did Ramanujan's mother agree to let him go abroad. Once at Cambridge, he produced some remarkable work. However, his methods were intuitive and inductive and sometimes incapable of being analyzed.

Ramanujan's interest in mathematics had been kindled by a book called "A Synopsis of Elementary Results in Pure and Applied Mathematics". This was a collection of thousands of mathematical results with the aim of helping English mathematics students pass the Tripos examination.

242 FAST FACT...

📖 **RAMANUJAN AND HARDY** worked on the formula for the number p("n") of partitions of a number "n". Ramanujan also worked on the hypergeometric series and the a function.

243 FAST FACT...

📖 **RAMANUJAN WAS** the first Indian to be allowed membership into the Royal Society of London.

244 FAST FACT...

📖 **RAMANUJAN'S PROOFS** of theorems on primary numbers were proved to be completely wrong.

245 PYTHAGORAS, FAMOUS GREEK MATHEMATICIAN 569 - 475 B.C.

PYTHAGORAS WAS BORN IN SAMOA, Greece, and is known as the first pure mathematician. He was given a good education by his father, who was a gem merchant. Pythagoras was a talented poet, musician, philosopher, and astronomer.

Pythagoras left for Egypt to gain more knowledge from Egyptian priests. He was responsible for influencing and creating several practices in Samoa, which had Egyptian roots. He considered the study of mathematics as the purification of the soul.

Pythagoras believed that mathematics form the base for everything and that geometry is at supreme for all mathematical lessons. He also believed that numbers had certain distinguishing and unique traits, characteristics, pros, and cons.

Pythagoras thought that some symbols had supernatural connotations. He also understood that the soul is immortal and a part of the brain.

In order to become pure, the soul moves from one living object to another, including humans and animals. This process of reincarnation of the soul is called transmigration.

Pythagoras studied different types of numbers like odd, even, triangular, and perfect numbers. This helped the future generations understand triangles, areas, polygons, etc.

The Pythagorean Theorem became an important pillar of mathematics and intrigued mathematicians for several centuries, which led them to develop more than 400 proofs to the theorem.

246 FAST FACT...

PYTHAGORAS FOUNDED a society which believed in the religious tenet of transmigration of souls and abstinence from eating beans as it was considered sinful!

247 FAST FACT...

WHEN PERSIA INVADED Egypt in 545 B.C., Pythagoras was held prisoner and sent to Babylon. He learnt a lot from the Magoi priests, who taught him sacred rites making him proficient in arithmetic and other Babylonian mathematics.

248 ARCHIMEDES (287 - 212 B.C.)

ARCHIMEDES WAS BORN in Syracuse, Sicily. He was a mathematician, philosopher, engineer, and astronomer. He was best known for his inventions, especially the "Archimedes screw", which was used for digging holes in the ground in a straight line. He studied the field of calculus by using smaller and smaller increments to measure the area of a circle and objects which had circular edges.

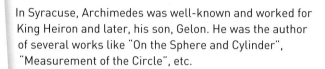

In Syracuse, Archimedes was well-known and worked for King Heiron and later, his son, Gelon. He was the author of several works like "On the Sphere and Cylinder", "Measurement of the Circle", etc.

Archimedes helped strengthen the fortress walls of Syracuse. He designed and constructed several war machines to defend the country. When the Romans attacked, they were surprised by the extensive preparations that had been made by Archimedes. All the moves of the enemy had been anticipated and counter attack measures helped destroy them.

250 FAST FACT...

ARCHIMEDES WAS KNOWN to be very absentminded. He would be so engrossed in his work that his students had to force him to take a shower! It was during his bath that he discovered the principle of buoyancy and solved the puzzle of the golden crown. When the concept struck Archimedes, he ran out on the streets shouting "Eureka!" meaning "I have found it!"

249 FAST FACT

KING HEIRON WANTED to check the authenticity of the gold used in a new crown made for him. He asked Archimedes to solve the puzzle without damaging the crown. He remained absorbed in solving this puzzle for a long time till he discovered the concept of buoyancy.

251 MARIA AGNESI

🎓 **MARIA AGNESI** was born in Milan in 1718. Her father was a professor of mathematics and encouraged his daughter to pursue her interests in the subject at a time when education in women was not considered important.

Maria spoke Greek, French, Hebrew, Spanish, and Latin very fluently, and helped educate her 20 siblings after their mother died.

The notes used by Maria to teach her brothers became her first publication. This publication was divided into two volumes and soon became important books in the fields of arithmetic, algebra, calculus, analytic geometry, etc.
The project took her 10 years to complete and was named "Analytic Institutions".

In recognition of her immense contribution, Pope Benedict XIV sent her a note accompanied by a gold medal and a gold wreath studded with precious stones. The Pope also appointed her to the chair of mathematics and natural philosophy at Bologna.

John Colson named the equation of a bell-shaped curve in mathematics, "Witch of Agnesi".

Maria Agnesi became a professor of Mathematics at a university and continued to teach till the death of her father in 1752. After his death, she left math and devoted the rest of her life to taking care of the sick and poor.

252 FAST FACT...

📖 **MARIA AGNESI** was known to sleep walk to her study and back to her bed in the night when a particular math problem troubled her. She would solve it in her sleep and have the solution waiting for her in the morning.

253 FAST FACT...

📖 **MARIA AGNESI** was deeply involved in taking care of the poor and sick and requested her father that her room be converted into a tiny hospital for this purpose!

254 SRINIVASA RAMANUJAN 1887 - 1920

SRINIVASA RAMANUJAN was a prodigy who pursued his interest in mathematics by himself. He was so interested in mathematics that he produced almost 4,000 theorems in number theory, algebra, and combinations. He wrote several letters to several mathematicians in different countries but none of them were responded to.

It was G.H. Hardy who recognized Ramanujan's mathematical ability and provided for his travel to Cambridge in 1913.

Hardy guided Ramanujan to use his intuitive and numerical abilities to provide unconventional and original solutions to complicated math problems that vexed professors in those days. Ramaujan's familiarity with the number theory was a rare gift in published journals of modular function theory.

Most of Ramanujan's work was done in collaboration with Hardy who was of the belief that if not for Ramanujan, most of the discoveries made would have been delayed by at least a century!

Ramanujan died at a very young age of 32, leaving a large part of his work unorganized and unpublished, which remains uninvestigated!

255 FAST FACT

ALTHOUGH RAMANUJAN had immense talent for mathematics, he was unable to pass his school examinations and could only get a job as a clerk in the city of Chennai.

256 FAST FACT...

HARDY VISITED RAMANUJAN at the hospital when he was sick and commented that he took taxi number 1729. Ramanujan instantly remarked that 1729 was an exceptional number where the smallest integer could be represented in two ways by the sum of two cubes.
1729=1 CUBE + 12 CUBE = 9 CUBE + 10 CUBE

257 NORBERT WEINER

AMERICAN MATHEMATICIAN NORBERT WEINER was born in 1894 in Columbia, Missouri. His father was a professor at Harvard and wanted to make his son a scholar.

Weiner was awarded a PhD by Harvard at the young age of 18. He studied philosophy, logic and mathematics, and became an instructor of mathematics at MIT in 1919. He collaborated with MIT on projects relating to anti-aircraft devices, which led to a number of contributions to the science of Cybernetics.

He became a visiting professor to a university in China and learnt Mandarin, because he believed that it was not possible to understand a nation without knowing its language. He traveled to Mexico and France, delivering a series of lectures.

Norbert Weiner published a lot of work on topics of logic, cybernetics, mathematical, and philosophical issues.

The concept of cybernetics, a transdisciplinary approach relevant to the study of various systems, has its roots in the "Cybernetics Group" created in 1943, where Weiner was an active participant.

Weiner was fond of life in the country and loved trekking and hiking. He was married to Marguerite Engelmann, who was a constant support for Weiner. She made sure that he got adequate privacy for his work and kept the intrusive media out.

258 FAST FACT

DESPITE BEING reminded several times by his wife about their change of address, the absentminded Weiner reached his old address. There, he met a little girl and asked her if she knew where the family had moved. To this, the little girl replied "Yes, daddy and mummy thought you would forget!"

259 FAST FACT...

WEINER'S WIFE would put pieces of paper with important messages in his pocket. During the day, he would scribble his ideas on those papers. If he was not satisfied with them, he would just throw the paper away, not realizing that his wife had written something important in them.

260 DAVID HILBERT

DAVID HILBERT was born in Königsberg, Germany, in 1862. He is considered to be one of the most influential mathematicians of the 20th century for his immense research in the fields of geometry, algebraic number field, etc.

Hilbert's work on calculus enabled him to invent the "Hilbert Space", an example of infinite-dimensional space, which is considered to be a key concept for functional analysis. Hilbert is credited with the development of the field of modern logic. He developed a new system of definitions of axioms in geometry, remodeling the work done by Euclid almost 2,000 years ago.

Hilbert was a young professor when he proved his "Finite Basis Theorem", which is an important aspect in general Algebra. At the International Mathematical Congress on 8th August, 1900, Hilbert made and presented the famous list of 23 unresolved problems, which continues to inspire and direct modern mathematicians.

Hilbert believed that young minds needed to be nurtured. He was also helpful in developing the careers of many students, who went on to become great mathematicians.

Hilbert is also known for the mathematical paradox called "Hilbert's Paradox of the Grand Hotel", which is about a hotel with an infinite number of rooms and infinite number of guests.

261 FAST FACT...

DAVID HILBERT was known diverting from his lecture topics. He would plan for one, but end up speaking about some other topic altogether!

262 FAST FACT...

WHEN HILBERT did not enjoy the company of a person visiting him at home, he would excuse himself politely and leave like he was the guest himself.

263 ÉVARISTE GALOIS

FRENCH MATHEMATICIAN ÉVARISTE GALOIS was one of the creators of group theory. He was killed in a duel at the age of 20.

Galois was also a strong political activist, which got him a prison stay for six months. One month after his release, he got involved in a duel, which was the result of his republican views.

Some say that the cause of the duel was more romantic. It has even been said that owing to Galois being an important republican, a conspiracy was set up to kill him.

A female agent in disguise goaded him and provoked him into the duel. Whatever be the reason of the duel, the result was fatal. He was shot in the stomach by his opponent.

Galois was left wounded and only taken to a hospital by a passing peasant. He refused the services of a priest and died a day later in the arms of his brother. It is also stated that he had a feeling he might be killed and therefore, he wrote down all his mathematical ideas. His writings are regarded as his scientific will and testament.

Galois scribbled in the margins that he had no time. In these writings, he coined the term "group" in its technical sense. These mathematical writings have been studied by mathematicians for years after his death.

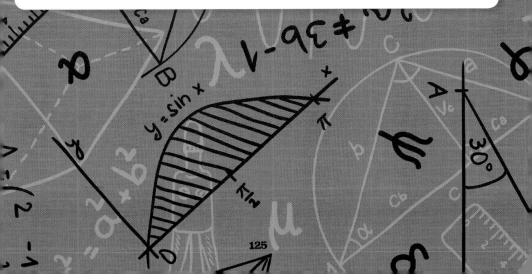

264 ARCHIMEDES AND PYTHAGORAS

🎓 **ARCHIMEDES IS ONE** of the famous ancient mathematicians known to mankind. He is famous for his laws governing floating bodies, and he also found a method for calculating large numbers, and developed an accurate estimate of "pi", the formula for volume of a sphere.

Apart from being a mathematician, Archimedes was a philosopher too. He made contributions to mathematics, philosophy, and engineering. He was killed in 212 B.C. at the age of 75 by a Roman soldier, probably owing to a misunderstanding. It is said that maybe he was wrongly identified, as his straight edge was taken to be a weapon. His killing was against orders as it had been declared during the siege of Syracuse that he should not be harmed. Archimedes was however, given a decent burial by the Romans.

Pythagoras is revered as a mathematician, musician, astronomer, and philosopher. He believed in mysticism and believed that some numbers were perfect and some were not. He even believed that numbers had personalities.

Pythagoras was shocked by discovering the non-finiteness of the root of 2. Pythagoreans had many strange beliefs and one of them was that beans should not be eaten. Pythagoras was very strict about the rules for his brotherhood.

The reason behind the death of Pythagoras is not known. However, there are speculations as to how he died. One involves Kylon, the son of a nobleman who wanted entry into the Pythagorean brotherhood, but was not ready or capable of observing the strict rules and discipline. He did not have the required intellect. Pythagoras refused to even meet him, turning him into an enemy.

Kylon grew vengeful and gave public speeches where he aggravated the people of Croton to believe that the Pythagoreans were manipulating them as if they were cattle.

The speeches angered the people so much that an angry mob attacked and set fire to the homes of the brotherhood. His students helped Pythagoras escape, but he soon came upon a field of beans. He was so enamored by the vision he saw before him that he stopped, thus allowing his persecutors to catch up with him and kill him.

265 LUDWIG BOLTZMANN

AUSTRIAN PHYSICIST AND PHILOSOPHER LUDWIG BOLTZMANN developed the link between the theory of thermodynamics and the kinetic theory of gases using the method of statistics. He was known as the man who trusted atoms. His theory of statistical mechanics was used to discover how large groups of atoms behaved. He explained how the properties of matter that were visible were dependent on the invisible properties of its atoms.

Boltzmann thought a lot on the field of philosophy and ethics, he was quoted saying, "Ethics must therefore ask when the individual may insist on his will and when he must subordinate it to that of others, in order that the existence of family, tribe or humanity as a whole and thereby of each individual is best promoted."

Boltzmann was a great teacher. His student Fritz Hasenöhrl said that Boltzmann had his heart in the right place as a teacher should. This is different from the requirements of a scientist, which are talent and intellect. A good teacher must understand those who are learning and he must be interested in their development. Hasenöhrl said that Boltzmann had earned the gratitude of many of his students.

However, his work on atomic physics was not given much recognition as he thought he deserved. It is extremely disheartening to know that such a wonderful man was prone to violent mood swings. Depressed by the lack of recognition, he committed suicide in 1906, when he was 60 years old. Sadly, 50 years after his death, his work gained the recognition it deserved.

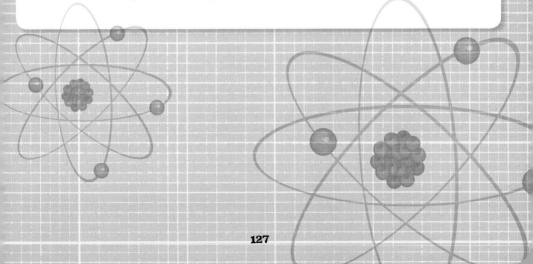

266 URYSOHN

🎓 **URYSOHN'S BEST** work was in the field of topology. He worked on the definitions of dimension and compactness.

Urysohn drowned at the age of 26 when he was swimming in the sea with his friend and colleague, Pavel Alexandrov. However, any articles published in French newspapers are untraceable.

Alexandrov, in his autobiography, has written about the death of Urysohn, which caused him great sorrow. He even provided pictures of where the death occurred. It seems the sea was very rough that day and even before they got into the water, a sense of apprehension had filled the two mathematicians.

Once in the sea, when they realized the danger they were in, they tried to swim back against the tide. However, spectators on the shore watched as they struggled helplessly, but Urysohn could not make it.

267 ALAN TURING

ALAN TURING is famous for his work in the field of computer science. He has achieved immense success in code breaking. He debated against the question "what does it mean for a task to be computable". Turing developed machines for performing computations which came to be known as the Touring machines.

Turing was under a lot of stress owing to his homosexuality, as it was illegal in Britain at the time. He was even blackmailed by his sexual partner. Taking recourse to the law against this blackmail, Turing was arrested.

However, he was let go with charges of indecent behavior. Being demoralized by this persecution, Turing committed suicide in 1954.

Turing committed suicide by eating an apple laced with cyanide. His death occurred under mysterious circumstances and some people believe that his death was an accident.

268 STANISLAW SAKS

SAKS WORKED with the theory of integration and measure. He was an inspiration to a number of young mathematicians. His political beliefs, however, were his undoing.

Saks was a part of the Polish army. Some say he was even part of the Polish underground. He was known to visit the Scottish Café. Saks moved to Lvov with the army.

However, when Lvov was invaded in 1941, Saks escaped to Poland. It is believed that he was still working hand in hand with the Polish underground movement.

In 1942, he was taken prisoner by the Gestapo. At the age of 44, he was executed.

269 FAST FACT

CONFINEMENT during the French Revolution proved to be a boon for Sophie Germain (and the mathematical world) because it was then that she took to studying geometry, which led to her important contribution to Fermat's last theorem.

270 DMITRI EGOROV

DMITRI EGOROV WORKED in the field of differential geometry, integral calculus and other areas of analysis. He developed a school to study real functions.

Egorov was also administrator of important organizations like the Institute for Mechanics and Mathematics (Moscow University) and the Moscow Mathematical Society. He was a person who was against persecution because of his religious beliefs. His opposition to religious persecution led to his dismissal from the Institute for Mechanics and Mathematics.

Egorov, however, continued as the administrator of the Moscow mathematical society, where he found support amongst his colleagues. However, external influences led to his release from this institution.

Egorov was dismissed and put under arrest. He continued his protests even in prison and went on hunger strike. He died of voluntary starvation in 1931. Where he died is a matter of debate. Opinions differ on whether he died in the hospital of the prison or in the house of a colleague.

271 FAST FACT

THE RELIGION of Jainism recognizes the custom of Santhara, wherein a person is allowed to liberate himself from this world by slowly abstaining from food. It is not suicide as it is not a result of depression or anger, but recognition of the failure of the body to serve its owner when death is inevitable.

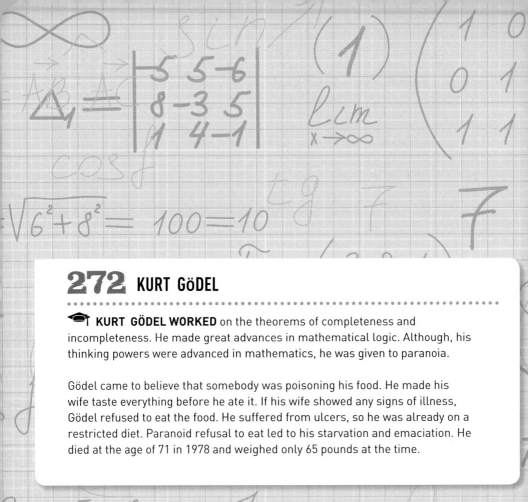

272 KURT GöDEL

🎓 **KURT GÖDEL WORKED** on the theorems of completeness and incompleteness. He made great advances in mathematical logic. Although, his thinking powers were advanced in mathematics, he was given to paranoia.

Gödel came to believe that somebody was poisoning his food. He made his wife taste everything before he ate it. If his wife showed any signs of illness, Gödel refused to eat the food. He suffered from ulcers, so he was already on a restricted diet. Paranoid refusal to eat led to his starvation and emaciation. He died at the age of 71 in 1978 and weighed only 65 pounds at the time.

273 ALEXANDER GROTHENDIECK

ALEXANDER GROTHENDIECK, a very important mathematician of the 20th century, was of mixed parentage. He was born to Russian-German parents in 1925. His father was killed during World War II.

The death of his father caused Grothendieck to become a staunch and vocal pacifist. He gave up mathematical work because of governmental and military influences on research in mathematics. He was awarded the Fields medal in 1966. He was even nominated for the Crawford prize in 1988.

Grothendieck went into complete retirement and refused to accept the medal. He made himself unavailable to the world in 1991. Today, he is barely known to the world.

274 MATHEMATICIANS WHO SUCCUMBED TO DISEASES CURABLE IN MODERN TIMES

THE WORLD LOST SOME remarkable mathematicians to diseases that are now curable. Andrey Markov died in 1897 at the age of 25 after being afflicted by tuberculosis.

Mathematician Niels Henrik Abel, being poor and underfed, contracted pneumonia and died at the age of 26. His mathematical work was not recognized and he never got a job as a mathematician in his lifetime. Abel never knew that a friend had managed to secure a job for him as a professor. A job would have meant the difference between life and death, allowing him to live a full life with a job and family.

Ramsey was a philosopher and economist. He also worked with combinatorics and logic/foundations. He suffered from a bad liver and was thus unable to work for more than a few hours a day. In spite of this, he did remarkable work.

However, when he was hospitalized and operated during an attack of jaundice, he succumbed to the disease in 1930 at the age of only 26. It seems all the more tragic that such deaths occurred even in 20th century.

275 FAST FACT...

EVEN INDIAN mathematician Ramanujan succumbed to a liver infection to which he was susceptible because of malnutrition in early life. He died at the age of 32 in 1920.

276 ISSAI SCHUR

🎓 **ISSAI SCHUR** is renowned for his contributions to Algebra. His theorem on monochromatic solutions to the equation "x+y=z" are very well-known.

Schur was Jewish and considered himself as much a German as any person of another race. However, he, like the other thousands of Jews, was made to leave Germany.

Schur did not have enough money even to pay the fee charged for Jews to leave Germany and had to borrow the same. He shifted to Palestine but was unable to find a job anywhere.

Schur tried extremely hard to find a job in USA, but was heartbroken when he was unable to do so. He died of a heart attack on the occasion of his 66th birthday.

277 NUMBERS AND CULTURES

SOME PEOPLE BELIEVE that numbers have positive and negative influences on the lives of humans. They believe that certain numbers bring good or bad luck to all humans. For example, the number 13 is considered unlucky in most cultures.

The number 13 is another number not favored by Christians as the 13th person on the table at the last supper of Jesus, Judas, was the one to betray him. In France, the number is considered extremely unlucky; it is said that Napoleon never had 13 guests at a dinner. In France, a "quartrozieme" or a paid guest is invited if there are 13 guests. Even the American president Herbert Hoover did not allow a meeting of exactly 13 people at the White house when he was in power. President Franklin Roosevelt invited his secretary to be the 14th guest if there were to be 13 otherwise.

Napoleon

Numerologists also believe that the date of birth of a person and even the number of letters in a person's name influence their personality and the fate of that person.

Franklin Roosevelt

People consider seven to be a lucky number. Some people even go to the extent of trying to give birth to their children on the 7th day of the month. The number seven has also been associated with power. The 7th son of a 7th son was believed to have supernatural powers. The 7th daughter of a 7th daughter was believed to have powers to heal and prophesize.

The number three is considered lucky by the Chinese, because the Chinese word for it sounds like the one for life. Christians believe in the holy trinity. Indians also believe in the trio of the creator, the preserver, and the destroyer.

The number four is considered lucky in most civilizations as this number is common in nature. However, the Chinese consider it unlucky as the word for four in Chinese sounds like the one for death. There is no Chinese aircraft bearing number four. The Chinese also consider the number nine to be unlucky. In China, hospitals and hotels may not have a fourth or ninth floor. In Russia, all odd numbers are considered lucky and all even ones unlucky.

278 THE NUMBER OF THE BEAST

🎓 **THE NUMBER 666** is considered by some people to be the number of the beast. The name was first used in the biblical book of Revelation, the 13th book of the Bible. It is also "the number of the numerologist".

However, the number has some properties which are of great interest to pure mathematicians. The number 666 is loved by numerologists like Edward Waring who loved to make up problems to express numbers in many different ways.

The number 666 can be broken up in several interesting ways. It is the sum of the squares of the first seven prime numbers, i.e., 2, 3, 5, 7, 11, 13 and 17. It is also the sum of the palindromic cubes (this means that it is the sum of the cubes of 1, 2, 3, 4, 5, 6, 5, 4, 3, 2, and 1).

The roulette wheel is marked from 1 to 36 and the sum of these numbers is 666. The total of the 1st six Roman numerals (I, V, X, L, C, D) is also 666.

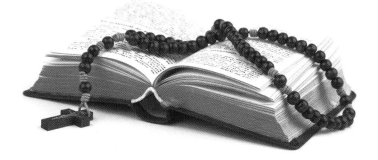

279 WOMEN MATHEMATICIANS IN ANCIENT TIMES

🎓 **IN ANCIENT TIMES**, women were considered intellectually inferior to men and were prohibited from learning or teaching fields of science, including mathematics.

Hypatia of Alexandria was murdered in 415 A.D. She was dragged from her carriage, stripped, and beaten to death. Her "crime" was that she was, to the knowledge of the people, the first woman in history to study astronomy, philosophy, and mathematics.

Hypatia was taught by her father, Theon. She had written several treatises on mathematics and this was believed to be sacrilege. Unfortunately, these treatises, for which Hypatia was murdered, were destroyed when the great Library of Alexandria was burnt.

280 NUMBER 786: ORIGIN

ARRANGEMENT OF LETTERS in Arabic is done in two ways. The first method, which is commonly used, is the alphabetical method. The second is the ordinal method, in which each letter is assigned an arithmetic value from 1 to 1,000.

It is believed that this arrangement began in the 3rd century of Hijrah. The system gained great importance and popularity in the Indian subcontinent.

The numbers "786" came to be a substituted for "Bismillah" or God. It is believed that the tradition of using 786 began in order to avoid writing the name of Allah on common papers.

As time passed, the number 786 became a symbol for Islam. People believe the holy number signified the blessings of the merciful Allah and brought luck and good fortune to his followers.

In the Indian subcontinent, the number's popularity goes beyond the beliefs of a single religion. Followers of other religions, too, uphold the piousness of this number.

281 FAST FACT...

"786" is the total value of the letters of "Bismillah al-Rahaman al-Rahim".

786

282 THE POWER OF 786

🎓 **SUPERSTITIONS CAN TRANSFORM** into beliefs overnight when a person feels that following a certain ritual is lucky for him. The symbolism of 786 gained prominence when people started believing that the use of it would bring good fortune to those who use it with a pure heart and sanctity.

Even today, the number 786 is used before starting a new or significant task. It is written on the top of documents, answer sheets, daily diaries, important memos, etc. It is believed that a new task must begin with Allah's blessings, which is why writing the numerical value of his name is auspicious.

Most houses and shops in Myanmar have the number 786 displayed in bold. It also shows that a particular shop, house or business belongs to a Muslim family.

Some people wear the divine number in the form of jewelry like chains, bracelets, etc.

Wedding cards have the number embossed on the covers to ensure peace and success for the completion of the wedding and a long, happy, and content life for the couple.

People try to get the lucky number on their vehicle license or number plates.

Many buy lottery tickets with numbers containing "786", expecting windfalls.

283 FAST FACT...

📖 **THE NUMBER 786** has been used in several films and documentaries to ensure that the films become box office blockbusters!

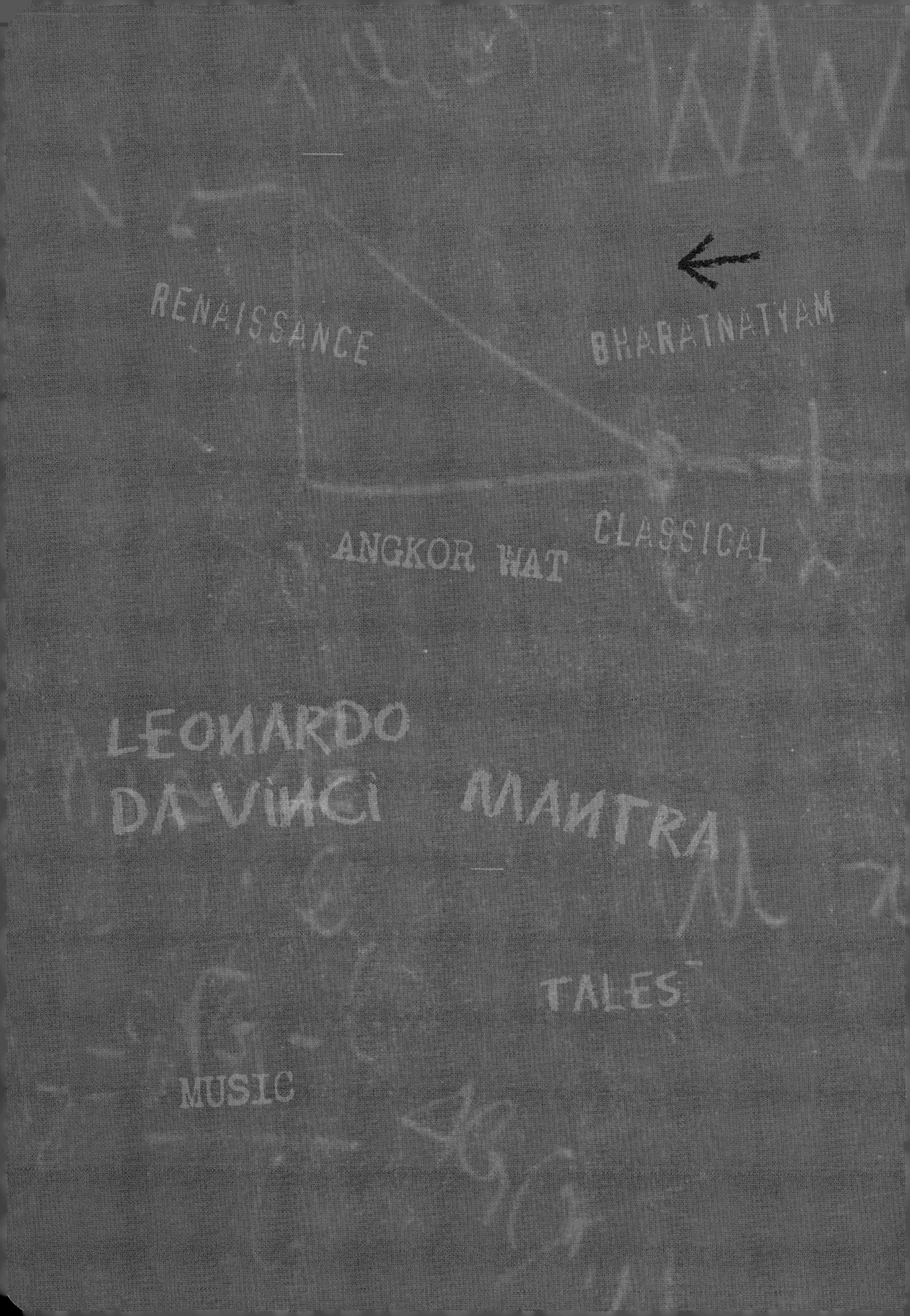

RENAISSANCE   BHARATNATYAM

ANGKOR WAT   CLASSICAL

LEONARDO
DA VINCI   MANTRA

TALES

MUSIC

DANCE
501 ISOMETRY

↓

Math and Culture

GOTHIC   MOZART

ZODIAC   RHYTHM
         ISLAMIC
              OM

284 THE MOZART EFFECT

MATH AND MUSIC are interrelated in many interesting ways. The link between mathematics and music has been researched in many areas like algebra, number theory, etc.

In 1993, a group of researchers from the University of California discovered that a set of college students improved their cognitive skills after listening to just 10 minutes of a Mozart sonata. Their IQ scores on spatial tests had improved by 8 to 10 points.

The researchers believed that listening to a particular kind of music stimulated the cerebral cortex of the brain that is responsible for motor control, speech control, listening, and memory skills. This in turn leads to a positive influence on a person's spatial temporal intelligence.

The tests were conducted using techniques like Positron Emission Tomography (PET) and functional magnetic resonance scanning. In related tests, the long term effects of music were studied in young children from three to four years who played the keyboard for six months. It was found that they performed better and achieved higher grades in mathematics than their counterparts who did not take keyboard lessons.

Most researchers in their tests have used Mozart's double piano sonata K448. Alfred Einstein called the sonata, "One of the most profound and mature of all Mozart's compositions."

285 FAST FACT...

JASON BROWN, a mathematician from the University of Dalhousie made use of a tool called "Fourier Transform" to resolve the puzzle about which instruments and notes actually made up the hugely popular opening chord of the Beatles' song A Hard Day's Night. (It wasn't just George Harrison's guitar skills!)

286 FAST FACT...

RESEARCHERS formed a group called the "Test Art" group which comprised of underachieving children who were given music and visual arts training consistently. They found that some students excelled in their class, even exceeding the marks of the usual high achievers.

287 LENGTH OF NOTES

🎓 **THE MOST IMPORTANT CONNECTION** between math and music is the length of notes. There are various lengths of notes that exist in music. With these different lengths, different types of music can be created.

What defines all of these notes and their lengths is a simple exercise of mathematical division. The basic note in music, called the whole note, generally has four beats.

Just like other mathematical concepts, there is a measure in music too. It is called the beat. Different genres of music follow different beats.

The rest of the notes are formed by dividing the whole note into two half notes, which are divided into quarter notes, which are further divided into eighth notes.

288 FAST FACT...

📖 **GOLDEN MEAN** is the ratio of two unequal parts of a segment (approx 0.61803). Mozart is known to have used the Golden mean in most of his sonatas. Another famous song having the Golden mean is "Hallelujah".

We can make use of a scale called the Pythagorean Scale to create all the notes from the harmonic series. This harmonic series helps us create perfect fifths.

For example, the fifth up from note C is note G, etc. Although, the Pythagorean Scale was mathematically perfect, there were drawbacks which were corrected by using the Tempered Scale invented by Andreas Werckmeister in the 1700s. This scale led the famous composer Bach to compose "The Well-Tempered Cavalier" in 1722, which uses each of the 24 keys.

289 DOING THE MATH DANCE

CHOREOGRAPHERS AND MATHEMATICIANS are coming together to develop the concept of a math dance. It is an innovative way where mathematical problem solving is integrated with complicated dance moves to create new forms of dances and develop new mathematics. These concepts can incorporate music and dance to help people who are not comfortable doing math in the classroom environment.

Geometry is the field of mathematics that is most closely related to dance. The shapes, patterns, symmetry, etc., help in determining the different angles and aspects of dance steps. Geometry could be used in choreographing a single dancer or a big group of dancers in various patterns and formations. The lines of their bodies could be used to create fluid and graceful movements that change with the music.

290 FAST FACT...

THE DECIMAL DANCE was introduced by some teachers to teach their class how to multiply decimals and count and set the decimal point in the correct place!

The theories of permutations and combinations can be applied to the concept of dance where a large number of people come together to perform. The steps could be choreographed based on the total number of people and the various ways in which they can be combined and partnered with each other.

Today, people are more open to combining unrelated streams of thinking to develop, explore, and create new fields which will benefit society in the coming years.

291 INDIAN CLASSICAL DANCE FORMS AND MATHEMATICS

JUST LIKE ANY OTHER DANCE FORM, the Indian classical dance form involves the weight of the dancer and the force of gravity. Experiments with gravity lead to intricate calculations and explore several expressions.

292 FAST FACT...

MOST INDIAN classical dance forms are narrative, sometimes depicting ancient epics through music and facial expressions!

293 FAST FACT...

CLASSICAL INDIAN DANCERS wear ankle straps with small bells ranging from 50 to 200 in number. The beat and count of each step that the dancers perform can be counted by the people in the audience!

Each movement is calculated as either going with the gravitational force or defying it gracefully. Indian classical dance forms utilize math in a complex, intriguing, and fascinating manner.

Every set of steps brings together complicated rhythmic patterns to intricate calculations involving mathematics. A form of classical dance uses ribbons to highlight triangles, parallelograms, etc.

The dance "Bharatnatyam" uses motions based on cardinal numbers three, four, five, seven, and nine to arrive at a figure of 32, which is the basic numerical structure of this ancient and beautiful dance form.

The various permutations and combinations of body postures are always in arrangements. For example, three is in arrangement with four, three and three add up to six, etc. The postures consist of angular nodes, formations of straight lines, and patterns of circular shapes.

Another classical Indian dance form, "Kathak", uses calculated mathematical beats that start with a slow pace and build up the tempo of the steps.

Sequences of steps are usually in multiples of three and the dancer must have an acute mathematical skill to finish perfectly on the last beat.

294. MATHEMATICS AND RHYTHM

🎓 **IF WE ANALYZE** the rhythm structures of different forms of music, we realize that these structures are completely mathematical.

Mathematicians develop algorithms and use tools like the Discreet Fourier Transform (DFT) and the Periodicity Transform to analyze frequencies and define the connection between math and rhythm.

Even when a single string of an instrument is plucked, it sets off a series of sound waves that can be converted into wave equations and studied in great detail.

Pitch and rhythm can be measured in cycles. Pitches are measured in hundreds and thousands of cycles per second while rhythms are measured in hundreds of cycles per minute. The tool DFT is used to identify prominent frequencies in a given signal.

The data obtained can be reduced and plotted on a graph for detailed analyses.

The data can be represented in a binary format using zero and one. Several pieces of popular songs have been converted into the binary format for clear understanding.

There is rhythm in nature everywhere, from tweeting birds to beating hearts, the ebb and flow of tides to the pulsation of stars. These phenomena are periodical in nature and these patterns can be measured with the help of detailed mathematical studies.

295. FAST FACT...

📖 **IN TIMES** when there were no pens, papers or other instruments for recording, teachers used to teach multiplication tables in a sing-song way because they believed that children would be able to memorize them better if they followed a consistent rhythm!

296. FAST FACT...

📖 **MINIMALISTIC MUSIC** involves an extensive repetition of simple melodies over slowly and progressively calculated changing harmonies.

297. FAST FACT

📖 **A METRONOME** is a device that produces regular beats at adjustable, precise, and measured intervals to help musicians maintain a steady tempo during their performance.

298   MUSIC AND MANTRA – ANALYZING THE "OM" CHANT

MANTRA IS A REPETITIVE CHANTING of ancient Sanskrit verses. Some of these verses are very small and some can be very long and complicated. The chants are done for a specific number of times.

The simplest yet most profound chant is considered the "OM" chant. Indian scientists use sophisticated mathematics to break down and study the vibrations of this traditional chant. Several journals have been published about this simple mantra.

The "OM" chant is small but has many variations and thus becomes more intriguing. It can be chanted in quick succession or pulses or can be drawn out in long intervals with stress on the end of the word for long periods.

The chanting of "OM" is called the peace mantra which has been used for thousands of years by Indian priests for prayers. Scientifically it has been explained that when one hears the "OM" sound the brain waves calm down from beta levels to alpha levels and finally down to gamma and delta levels.

Every syllable of the mantra when dissected is found to produce tonal frequencies that resonate with great energy. Scientists believe that the pitch emanated from the chanting of "OM" is identical to the "Sound of the universe".

299   FAST FACT

MATHEMATICIANS ARE WORKING on studying some mantras which are supposed to have healing powers. These mantras have been in existence for thousands of years and once decoded, these healing mantras can prove to be beneficial.

300 ARTISTS USE MATHEMATICAL TOOLS – ISOMETRY

THE ISOMETRY OF A PLANE is a linear transformation and includes the rotation, translation, reflection, and glide of geometric figures. It is an important tool used by artists to ensure that their work is symmetrical and visually appealing.

Artists use the concept of symmetry to create a single motif or ornament that has uniformity. This single motif can be built up to create complicated patterns. Mathematicians like Hermann Weyl have been instrumental in teaching artists the concept of symmetry and patterns.

Artists use mathematical concepts to develop "polyhedrons" or solids which have multiple flat surfaces, like pyramids, prisms. Polyhedra have inspired artists and mathematicians to create a number of objects from ingenuous puzzles to intriguing pyramids.

Artists use the concept of polyhedron for assembling and dissecting models and enabling unexpected transformation of shapes.

Mathematical theories are increasingly used in the seemingly simple art of "Origami." Artists use geometric axioms to explore all the new shapes that can be constructed by folding a piece of paper.

301 FAST FACT...

M. C. ESCHER was an artist whose works include a considerable use of Math. He is famous for his lithograph print called "Waterfall", which has been described as a visual paradox.

302 RENAISSANCE ARTS AND MATH – PERSPECTIVE CONSTRUCTION

THE RENAISSANCE is known to have been a period when art went through a lot of transformation. Artists incorporated a new quantitative approach to perceive and present their creations to the world. Attention was paid to the realistic measurements of distance, shape, etc., in paintings and other art forms.

Distances were measured from the "eye" to the picture-plane and use of that distance in a plane-geometric construction. For example, Leon Battista Alberti's, Della Pittura (1435), explained in detail how a simple thing like a pavement must be drawn correctly and in perspective.

Artists like LodovicoCardi (also known as Cigoli) were specially trained by mathematicians like Galileo Galilei who were masters of perspective construction.

Art during the renaissance used abstract geometrical concepts and made them into important working knowledge for all artists. Art became an amalgamation of perfect theory of measurements with an imperfect medium.

Galileo Galilei is credited for introducing the correct interpretation of patterns of light and dark and subsequently incorporating this knowledge into shading works of art. This technique was called "chiaroscuro".

In 1984, Samuel Edgerton writes "I shall argue that we have here a clear case of cause and effect between the practice of Italian renaissance art and the development of modern experimental science."

303 FAST FACT...

IT IS A MYTH to think that some people have a "math mind" and others don't. All it takes is the right attitude and confidence that is needed to succeed in math.

304 LEONARDO DA VINCI'S ART AND MATHEMATICS

🎓 **LEONARDO DA VINCI** lived from 1452 to 1519 in Italy. He was an accomplished mathematician, painter, architect, and engineer. He was known not only for his work in art but also his ideas in generating mechanical objects.

Da Vinci is known to have used the technique of seeing with one eye at a time to gain a different impression of depth and then use gradual and calculated diminution in the intensity of colors, depending on the distance.

Da Vinci had a technique where he observed objects coming towards his eyes, then placed a paper (just like a graph paper) between his eyes and the object to get the important points and achieve realistic results. This technique was central to the fields of science because of the absence of photography; everything that had to be recorded had to be drawn.

Da Vinci believed that nature can be best described through mathematics. He used his inventive imagination, frequent experimentation, and knowledge of technology and mathematics to create art. Da Vinci is considered to be one of the few geniuses who have been an inspiration to several other talented minds.

Leonardo Da Vinci knew that mathematics was an essential aspect of paintings and he believed that anybody who was not interested in mathematics did not need to begin reading his works.

305 FAST FACT

📖 **DA VINCI** wrote with his left hand in a way that it was laterally inverted! Most people believe that this was done to preserve the secrecy of his work.

306 THE ANCIENT PYRAMIDS OF EGYPT AND MATHEMATICS

THE RUINS of the majestic pyramids in Egypt demonstrate that the people who conceived, designed, and constructed them were not just architects but accomplished mathematicians who had high regards for the cosmos. Great attention was paid to details of spatial co-ordinates while building the pyramids at Giza.

The pyramids faces are built precisely along the lines of the four cardinal directions, leaving very little place for error. The pyramids match the Orion's belt, almost perfectly indicating that the construction engineers were proficient in trigonometry.

The Great Pyramids of Giza was built with thousands of stones that were measured, cut, and polished to perfection, then placed together in such a compact way that even a razor thin blade could not enter the crevices. This fact continues to puzzle researchers even to this day!

Research shows us that the construction engineers of these pyramids knew the use of pi and phi or the golden mean and they had great knowledge about triangles, thousands of years before Pythagoras lived.

Research also reveals that designers knew the precise dimensions of the Earth's spherical shape and accurately charted the occurrence of events like equinoxes and solstices.

307 FAST FACT...

RESEARCHERS from Czechoslovakia and USA conducted a geometrical study of the pyramids and found that the form of the pyramids magically kept food from decaying, enabled plants to germinate faster, and hastened the process of healing wounds!

308 FAST FACT...

IT IS ESTIMATED that the weight of the Great Pyramid is 6,648,000 tons!

309 THE TEMPLE OF ANGKOR WAT AND MATHEMATICAL ASTRONOMY

THE MONUMENT studied in minute detail for its cosmological significance is the temple of Angkor Wat in Cambodia. It was built on the basis of Indian architecture and had deep connections to mathematical astronomy.

The temple was built for observations of lunar and solar phenomena. Researchers decoded that the various dimensions within the temple were based on numerical and calendric significance.

Angkor Wat is considered to be a masterpiece of perfect proportions and intricate sculptures built with fine precision, depicting scenes of Indian mythology.

The altars inside the temple are perfect geometrical figures like squares, circles, and semicircles. Measurements are made in a perfect ratio at the level of the universe and the individual in order to enable humans to achieve spiritual transformation.

A special significance of number 108 was found in several areas of the temple. The number 108 represents the distance of the Earth from the Sun and the difference in the Sun and moon's diameters respectively.

An important point noticed about Angkor Wat was that the height of the temple is exactly double its width and the height of the foundation is exactly equal to a third of the temple's height. These perfect measurements indicate that brilliant mathematicians played an important role in building this architectural marvel.

310 FAST FACT

AROUND the main structure of the temple is a wide moat and a boundary wall that almost 15 feet tall, 3,360 feet long, and 2,630 feet wide!

311 BABYLONIANS AND THE PYTHAGORAS THEOREM

🎓 **GOTHIC ARCHITECTURE** is usually characterized by its pointed arches, vaulted ceilings, stain glass windows, etc. The interiors of a Gothic church had tall pointed arcades and heavy structural features. This style of architecture was prevalent (majorly in Europe) from the 12th to the 16th century.

Gothic churches and cathedrals have a grace that powerfully appeals to us in many ways. The elegance of such structures comes from the fact that these churches were engineering marvels. The barrel vaults, the flying buttresses, the tall and elegant spires, and flights of perfectly cut stone staircases were all examples of ideas born out of mathematics.

One can only imagine the detailed mathematical calculations gone in to make the cathedral geometrically well-proportioned, from labyrinths to perfectly placed tiles to structures that support the ceilings.

Architects only had a few mathematical instruments to create these marvelous Gothic buildings. Engineers overcame the problem of fire by eliminating wooden ceilings and increasing their height.

In order to ensure that the tall cathedral structure did not collapse inwards, calculated flying buttresses were built. Large stain glass windows provided light to the interiors.

312 FAST FACT...

📖 **THE WORLD'S** largest Gothic cathedral, The Cathedral of St. John the Divine, is situated in New York City. It is 601 feet long and 146 feet wide.

313 FAST FACT...

📖 **THE LARGEST** stain glass window at the Kennedy International Airport's American Airlines Terminal measures 300 feet in length and is 23 feet in height.

314 ISLAMIC ARCHITECTURE AND MATHEMATICS

🎓 **ISLAMIC ARCHITECTURE** can be classified by its symmetrical domes and beautifully intricate patterns and carvings. Recent discoveries in the field of Islamic architecture suggest that Muslim architects and engineers used mathematics way more than just to produce decorative patterns.

We can deduce this by studying the mosque called "Divrigi Ulu" in Sivan, Turkey. The mosque has beautiful geometric designs, indicating that its engineers had deep knowledge of geometry and symmetry.

A unique aspect of this mosque is that silhouettes of the carvings appear on the outer walls of this mosque at different times of the day.

Four silhouettes appear on the walls facing different directions. The first three resemble a man looking straight ahead, reading a book and praying. The last one is that of a female figure who is also praying.

315 FAST FACT...

📖 **THE TAJ MAHAL** in Agra, India, is probably the world's most beautiful example of Muslim architecture, characterized by reflections, symmetries, octagonal, and hexagonal layouts. Even the structure's gardens are constructed with precision and flawless tiling.

These amazing features lead us to believe that mathematicians, astronomers, and artists must have observed the positions of the Sun for long periods of time. It is obvious that only after careful and precise calculations would the engineers have begun their work; otherwise the shadows formed would not be that precise.

316 MODERN ARCHITECTURE AND MATHEMATICS

MODERN DAY architects have the benefits of planning and designing with the help of computers and other advanced digital tools. Virtual models of the building help make construction and planning easy.

Architects can explore different mathematical tools to create new types of surfaces and textures. The most important innovation in the field of architecture is called "parametric modeling".

317 FAST FACT

MATH requires both logic and intuition. Intuition is the cornerstone for doing math productively.

Parametric modeling allows the architects and engineers to make modifications to the computer models of the buildings without having to completely change the original plan. Calculations and geometrical features can be changed several times till an appropriate result is achieved.

Architects and engineers have used this modern concept in constructing the iconic "Gherkin", which stands out beautifully on the London skyline.

Another example of modern architecture is the London City Hall, which is built like a sphere to minimize surface area for energy efficiency.

Mathematicians believe that amongst all solid shapes, the sphere has the least surface area when compared to volume.

318 SCIENCE OF MATHEMATICAL ASTRONOMY

MATHEMATICAL ASTRONOMY is described as a scientific study of the various aspects of the universe using mathematical tools and equations.

Astronomy and math have a strong link as all astronomers, from early civilizations to this date, need to be good mathematicians.

The Mayans, Mesopotamians, and Babylonians had accomplished mathematicians who studied the movements of heavenly bodies to calculate the occurrence of celestial events.

Present day astronomers use mathematical fields like algebra, geometry, trigonometry, and logic to understand and study space.

In the 1700s, William and Caroline Hershel developed modern mathematical methods to study astronomy, which led to the discovery of the planet Uranus.

In the 1800s, Urbain Jean Joseph Le Verrier used mathematical equations to determine the existence of the planet Neptune.

Verrier's calculations were so precise that astronomers from other countries could spot the planet easily using his equations.

319 FAST FACT...
GALILEO GALILEI was an Italian astronomer who had enrolled in the University of Pisa to study medicine. He never finished the course and went on to become an astronomer, mathematician, physicist and inventor of the telescope!

320 FAST FACT...
A PARSEC is a mathematical measure of length in astronomy. One parsec is equal to 3.26 light years.

The most important use of mathematics in astronomy is for calculating the distance of a celestial object from Earth along with calculating the distance between two or more heavenly bodies.

321 APPARENT AND ABSOLUTE MAGNITUDES AND LUMINOSITY OF A STAR

CHILDREN often wonder how stars glow in the night sky. Early Greeks described stars by their level of brightness. But this had its limitations, as they could only describe what they saw from the Earth.

The scale compiled by them only showed the apparent brightness of the celestial body and not the actual brightness if it were viewed from a close range.

The luminosity of a star is dependent on its temperature and radius or surface area. A hot star (with high surface temperature) would be more luminous than a cold star (with low surface temperature). Similarly, a bigger star would be more luminous than a smaller one.

Mathematicians developed a formula to determine how bright a star actually was if we were to observe it from a defined fixed distance. For this purpose, calculations have to be made to find the distance of the star from the Earth and what its magnitude would be if it is placed at 10 parsecs from the Earth. For example, the star "Rigel" appears brighter than the star "Deneb" in the sky.

Scientists have found that both have the same absolute magnitude, but Deneb is further away than Rigel, because of which there is a difference in brightness.

322 FAST FACT

THE SPACE probe Voyager 1 was 0.0006 parsecs from the Earth as of May 2013. It took 35 years for the space probe to cover this distance!

323 GEOCENTRIC THEORY

🎓 **EARLY GREEK MATHEMATICIANS** observed the sky and believed that the Earth was at the center of the solar system. They believed that all the planets moved around the Earth in a clockwise direction.

It was easy for the Greeks to believe that the Earth was stationary as no one could feel any movement. Since they saw the stars moving in a circular pattern everyday, they were further convinced that these stars were revolving around the Earth, leading to the theory called the "Earth Centric" or "Geocentric theory".

The knowledge of the Greeks on subjects like trigonometry led them to conclude that faster moving objects are closer to Earth and slower moving celestial bodies are further away.

Claudius Ptolemy wrote many volumes on the geocentric theory in 140 A.D., which was accepted by the Roman Catholic Church. Anyone who opposed this belief was punished with house arrest!

It was the work done by Galileo, Copernicus, and Kepler that formulated the heliocentric theory which is currently followed. Their views were initially refused by the Church and Galileo not only had to withdraw his work, but also spend the last years of his life in jail!

Kepler

324 FAST FACT...

📖 **KEPLER** believed that it was the moon which caused tides in the ocean. However, Galileo believed that tides were caused by the Earth's rotation!

325 MATHEMATICS AND EARLY INSTRUMENTS USED IN ASTRONOMY

ONE OF THE FIRST INSTRUMENTS used in Astronomy was the astrolabe, which was a circular disc marked with degrees along its rim. This instrument was used to measure the altitude of stars and planets.

The "quadrant" was another device used for measuring the positions of celestial objects. It was a 90° arc like a mathematical protractor with a pointer for sighting stars.

A sextant was an instrument used by sailors to navigate the sea by observing the stars. These instruments were used to measure stellar positions before the telescope was invented.

The invention of the telescope transformed the science of astronomy. There is a constant study happening for the development of better telescopes with several multi-faceted mirrors that are controlled by computers.

With the development of better instruments and electronic methods, most celestial bodies can be studied by analyzing the X-rays and gamma rays emitted by them.

Modern day scientists and mathematicians study the universe by analyzing the electromagnetic spectrum and studying the UV, infrared, and microwave wavelengths.

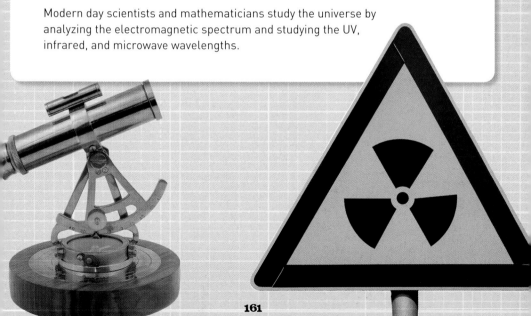

326 MATHEMATICS AND ZODIAC

🎓 **WHEN WE LOOK** up to the sky, we do not see a complete circle. However, if the sky is measured mathematically, we derive a circle of 360°. This circle, when divided into 12 equal and meaningful sections, gives us the 12 signs of the zodiac.

The division of the sky is done by carefully calculating the movement of the Sun over the course of one complete moon cycle (which is from full moon to new moon and back). These very precise 12 divisions are 30° each.

Astrologers study these 12 signs in great detail. As the Sun takes its elliptical path, it enters each of the 12 houses of the zodiac. The individual born on a certain date takes the sign of the house that the Sun is in during that moment.

Using simple geometry, the zodiacs can be further divided using a triangle, square, and hexagon. Astrologers divide the signs using the triangle as the natural elements of fire, Earth, air, and water.

When division according to the square is used, the signs get divided into cardinal, fixed, and mutable signs. When the hexagon division is used, these signs get segregated into positive and negative.

327 FAST FACT...

📖 **IN ASTROLOGY**, the number 11 is considered a magical number as it is supposed to strike a balance of emotion, thoughts, and spirits!

328 MATH AND NATAL CHARTS

🎓 **A PERSON'S NATAL CHART** is like a mathematical graph or a roadmap of his/her physical, emotional, spiritual, and intellectual journey of life. An astrologer first mathematically establishes the natal chart of a person.

The Astrologer then goes on to interpret it by analyzing all the planetary positions just like a scientist would analyze the recorded observations for his work.

The Astrologer makes deductions by determining the position of the planets within specific houses and the mathematical relationship of one planet with another. The astrologer then makes recommendations regarding these relationships and a positive use of the energies formed. Thus a natal chart is a blue print of our potential and can transform our lives.

329 MATHEMATICS AND "MUHURAT" (AUSPICIOUS TIME)

🎓 **IN ALMOST** all parts of India, precise calculations are made to figure out the most auspicious times of the day when important events like marriages and house warming ceremonies can take place.

Conversely, priests warn of inauspicious times when all these events must be avoided. It is a common belief in the country that events which take place during good "muhurats" are successful and fruitful.

The knowledge of time is called "Kaal Gyan" which has several sections related to time. The knowledge of one's actions during a certain time is expressed in "muhurat"; this is, in turn, recorded through math calculations. It is believed that people, time, and place are linked by planets, stars and zodiac, and there is no scope for miscalculation or imagination.

According to the Indian system, there are 27 stars in our constellation, 12 zodiac signs and 9 planets. Detailed calculations based on the position of these stars helps astrologers predict events with precision.

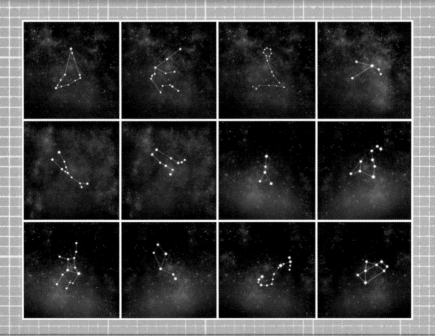

330 FAST FACT

 **THE CHINESE** developed a chart which can help a couple calculate and predict the sex of their child!

331   A TALE OF TWO LUMBERJACKS

🎓 **AN ODD COMPETITION** was held in the forest of a city. The forest was thick and impenetrable, and people wanted to clear a patch of it. Toby and Jacob were two lumberjacks of the same size. They decided that this was an opportunity to earn some more money, which was a cash prize of USD 15000. However, the rules of the competition strictly stated that only one axe could be utilized by each contestant.

When the competition began, Toby began working on his patch with full vigor, keeping his eyes and ears wide open to listen to his competition. He was surprised and pleased to learn that every 10 minutes, Jacob took a break. Happy at his new discovery, Toby worked faster and with passion.

However, in the evening, when the number of trees were counted, Toby was astonished to find that Jacob had managed to chop down more trees than he could. When Toby asked Jacob how he managed such a feat with a break every 10 minutes. Jacob simply replied saying that he took those breaks to sharpen his axe so that he would need fewer chops to bring down the tree. This obviously goes to show that mathematics and common sense are joint at the hip.

332. COUNTING THE CHRISTMAS ORNAMENTS

CHRISTMAS TIME BROUGHT out a different kind of happiness to six year old Annie's face. This year was no different. Annie was excited that she was in charge of pulling out the ornaments that went on the tree, and her three year old brother Andy had decided to help. Andy and Annie loved each other to bits and never did anything without the other.

Annie remembered how her parents would carry Andy and her to the couch and let them watch while they decorated the tree. Annie's mother would count each ornament while placing it on the tree, making Annie recite the numbers behind her in a sing song voice.

Annie had learnt her basic counting by placing large and tiny decorations on their stunning tree year after year. This year, Annie decided that it was Andy's turn to learn.

As Annie opened the box, Andy sang out in a familiar tune "I want to open...I want to place...one...two...three!" Annie smiled proudly at her brother...he had started to learn how to count!

333. GOOD INDUCTION, BAD INDUCTION

A SCIENTIST was once conducting an experiment in his laboratory, for which he needed a jar filled with 100 fleas. He carefully picked up a flea and placed it on the table. He then ordered the flea to jump. The flea obeyed and was placed into an empty jar. Another flea was picked up and ordered to do the same; it followed his orders and was placed into the jar with its counterpart.

The scientist continued doing this till all the fleas were in the other jar. Once again, he picked up one flea from the now filled jar and pulled off its hind legs. He then directed it to jump, but the flea did not obey and was placed in another empty jar. He followed this process till the other jar was full of fleas that refused to obey his command.

In his research findings, the scientist then wrote about how a flea cannot hear if its hind legs are cut off.

334 TWO PLUS TWO

🎓 **SIR FRANCIS GALLON**, English scientist and explorer, spoke about the primitive Damaras of Africa and their level of mathematical comprehension.

Gallon said that in the process of a barter, one sheep was sold for two sticks of tobacco. However, a Caucasian trader once offered a tribesman four sticks of tobacco for two sheep.

This was not comprehended by the tribesman easily and a less complicated solution was found. One sheep was first given in exchange of two tobacco sticks and the same was followed for the other. On receiving the same amount, the tribesman believed the Caucasian to be some sort of wizard.

However, the Damaras should not be considered unintelligent. They knew exactly what the amount of a flock of sheep or a herd of oxen would be. They could easily miss an individual but recognize the faces of animals easily. This form of intelligence would be very rudimentary for us. However, we must understand that it is accurate and intense observation and it would be much more complicated to develop than that which concerns counting.

335 THE CROW WHO COULD COUNT

🎓 **A NOBLEMAN** in France once grew annoyed and wary of a stubborn crow who had decided to build its nest in the man's house. The man repeatedly ordered his men to put an end to the crow, but they failed. Every time, the men tried to shoot the crow, it would fly away to a distant tree and keep a watchful eye on its nest.

The nobleman grew tired of the crow and organized a contest to help him get rid of the menace. Soon, men began to try their luck. They devised plans where two men would go into the house, one would come out while the other waited inside patiently for the crow to return and then shoot the crow.

However, the crow was smart and did not fall for the trap. Even when four men were sent in and three returned, the crow kepts its distance. When a party of five men was sent and four returned, the crow grew confused and could not tell the difference. It flew back to its nest, with one man waiting patiently for its arrival.

The conclusion of this story is unknown yet; but let us believe that the crow was shown mercy for its brilliant skills in mathematics.

336 THE DECEITFUL GOLDSMITH

🎓 **ARCHIMEDES IS KNOWN** to have been capable of strong mental application. A story that is not unheard of is the one with King Hiero.

King Hiero, desired a crown of gold, and gave the weight of the metal to a goldsmith, along with instructions.

After a period of time the crown was finished and handed over to the king. The crown was beautiful and marvelously crafted, the kind suspected some fraud. The king believed that silver had been mixed with gold while making the crown. Not wanting to break apart the beautiful work. The king asked Archimedes for a solution. Archimedes was baffled himself.

One day, when Archimedes was having a bath, he came across the solution, discovering the first law of hydrostatics.

In his excitement, he forgot to put clothes on and ran out into the street shouting, "Eureka, eureka" ("I have found it, I have found it").

The famous first law of hydrostatics appeared later as Proposition seven in his first book called "On Floating Bodies."

The law states that, "Everybody immersed in a fluid is buoyed up by a force equal to the weight of the displaced fluid."

337 FAST FACT

📖 **RESEARCH SHOWS** that both men and women are equal in their math skills. Women tend to give up more easily sometimes.

338 FAST FACT...

📖 **IT IS INCORRECT** to think that one needs a good memory to solve math problems. If the concepts make sense, then remembering them is no problem.

339 FAST FACT...

📖 **THERE IS NOTHING** wrong with counting on fingers. On the contrary, it shows a better understanding of arithmetic.

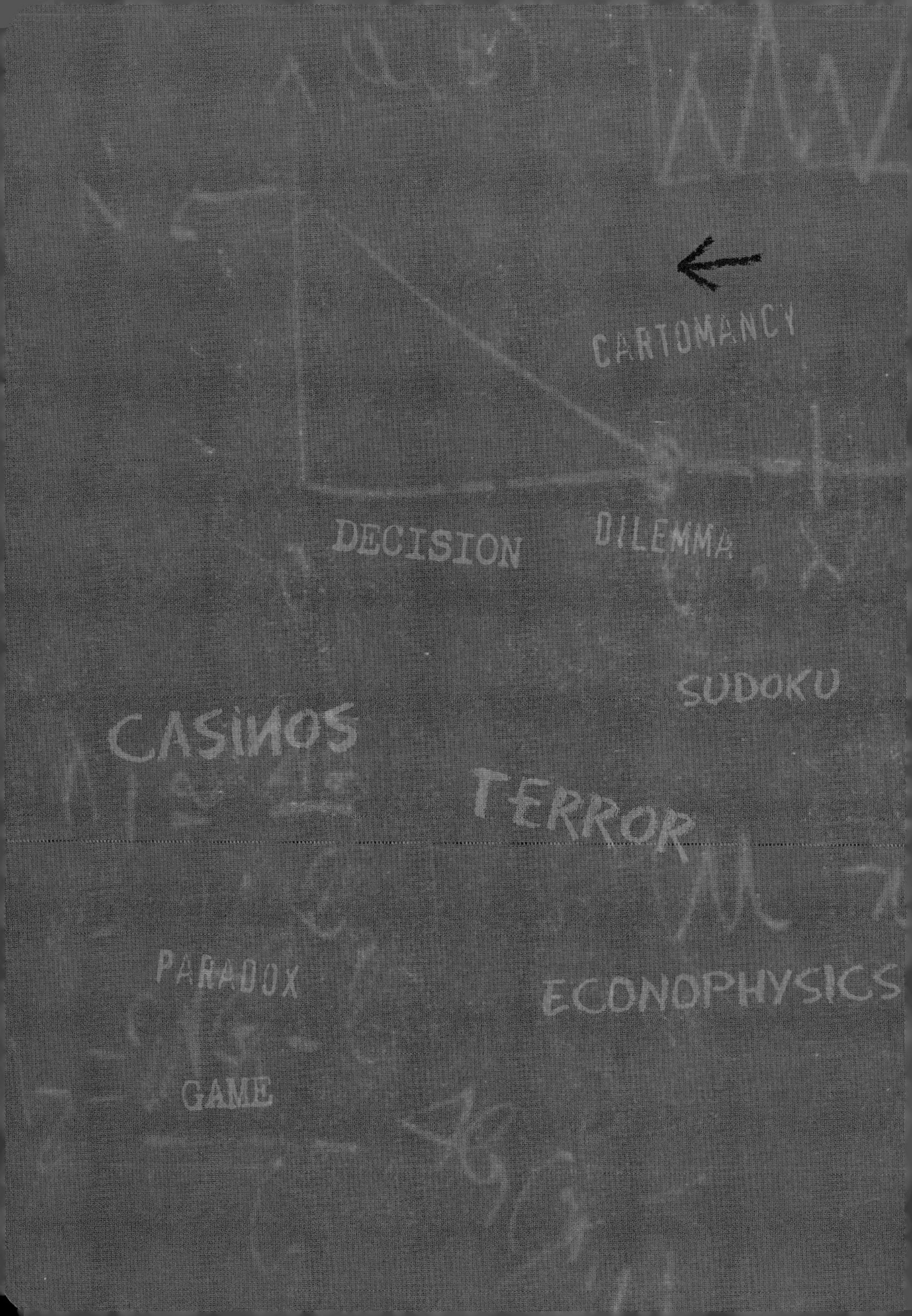

CARTOMANCY ←

DECISION   DILEMMA

                    SUDOKU

CASINOS

            TERROR

PARADOX   ECONOPHYSICS

GAME

501 Math and Probability

340 MATHEMATIC OF DECISION MAKING

🎓 **WE ALL MAKE** decisions in our everyday lives. Sometimes, we make choices based on prior information, at other times we rely on our instincts. This decision making process involves the use of math to a very large extent, because making the correct decision involves an element of some risk and uncertainty.

Mathematicians developed a theory called the "Decision Theory", which teaches how to make the best choice given the various options and consequences thereof. For example, a doctor may have to make a choice between several treatment options for his patient, knowing that there is very little scope for experimentation to check which treatment would be the best.

Another example would be, a business management team choosing between several advertising campaigns and selecting the one which would be successful for their business. All these decisions have to be made by calculating an expected value.

Once we calculate the expected value of the options available to us, we can arrive at a decision by choosing the most rational option that gives us the best returns or rewards.

341 FAST FACT...

📖 **HEURISTICS ARE MENTAL SHORTCUTS** that people use to make decisions quickly. Sometimes heuristics can be helpful, but at other times they can lead to biases.

342 ESTIMATION AND MATHEMATICS

🎓 **ESTIMATION IS A HELPFUL TOOL IN MATHEMATICS** for calculating the periods of time, amount of money, distances, etc. Estimation is also called an "educated guess" or an "intelligent prediction" of the calculation's outcome.

People find estimation to be a preferable method for data analysis. Estimation strategies are taught in schools so that children understand the use of averages, comparison of two quantities, etc.

Estimation is a mix of content and process. Everyday, in our lives, we need these skills to focus on the rationality of an answer and to know if the answer values fall in a range of tolerance. Statisticians often refer to the term "eyeballing" the data, which refers to the process where values are derived from a set of data. This is known as the estimation of probability.

An important part of mathematical estimation is to recognize the limitations of estimations and appropriately assess the error resulting from this approximation. This error is known as variance.

Estimation theory is applied in various fields like interpretation of scientific experiments, opinion polls, project management, and quality control.

343 MATHEMATICAL PROBABILITY

PROBABILITY IS THE CHANCE OR LIKELIHOOD that a certain event may occur. We can measure this probability in mathematical terms in the form of a ratio. The ratio is calculated by giving values to the event occurring or not occurring.

The fractions that denote the number of ways an event will happen or fail to happen divided by the total number of possible outcomes is called the mathematical probability of that event.

Tossing a coin, throwing dice, etc., are some of the events where mathematicians use the probability theory. They can also compute if the events are mutually exclusive (meaning the events cannot occur simultaneously), independent, dependent, etc.

Weather reports, election results, product surveys, probability, estimations, and forecasts have become an important stream of mathematics today. We read and understand data, make inferences, formulate hypothesis, and evaluate arguments based on our knowledge of probability.

344 FAST FACT...

📖 **FLORENCE NIGHTINGALE** is considered to be a pioneer of estimation and the statistical inference of public health. She collected and analyzed public records of cholera cases in London during the cholera epidemic of 1854 in order to combat and control it.

345 MATHEMATICAL PROBABILITY AND PARADOXES

🎓 **MATH IS A LANGUAGE THAT IS SPOKEN AROUND THE WORLD** because it is a language of logic. However, this system of proof and logic has certain paradoxes and contradictions which have troubled mathematicians for centuries.

Some paradoxes are just logic tricks; others trouble mathematicians and require them to use brilliant thinking and creative imagination to resolve them. The study of paradoxes is a useful tool that plays an important role in mastering a new stream of mathematics. This can be challenging and has a lot of real life application involved.

Lewis Carrol's "Urn" is a famous paradox. An urn contains a single ball which is either black or white with equal probabilities. A white ball is then added to the urn and one ball is taken out randomly. That ball turns out to be white. How must we determine if the ball in the urn is white too? The paradox is that the answer is not ½.

Another important paradox is called the "Elevator and the Paranoid". An office building has one elevator that spends the same amount of time on each floor. One man's office is on a higher floor and he has to wait for the elevator to take him to the top floor to his boss's office. He believes that the first elevator to stop on his floor is the one that is going down more frequently than going up. Is he simply paranoid or is there a rational explanation to it?

346 UNDERSTANDING ODDS

ODDS ARE DEFINED AS THE "amounts staked by the parties to a bet based on the expected probability." Odds represent the ratio of the mathematical probability of an event happening. Odds in common usage are spoken as "odds for" or "odds against" an event's occurrence.

Odds are most commonly used in the area of horse racing. This sport uses the concept of "pari-mutuel odds" to calculate the payouts on the races. This system was developed in Paris to make sure that a fair system is provided for users, and they get maximum profits, irrespective of the race's outcome. This system is based on a "betting pool table", which displays pari-mutuel odds, the percentage of pool, and the payouts.

This system is designed to ensure that the track makes a profit no matter what the outcome of the race is. The term "take out" is a percentage of the total pool that is taken by the track before any payoffs are made. This involves careful mathematical calculations, because there is a great deal of money involved in the process. The take out varies from place to place.

347 FAST FACT...

📖 **THE WEIGHT OF JOCKEYS** plays a crucial role. In some horse racing derbies, the weight of the jockey and his equipment cannot be more than 126 lbs. Therefore, most jockeys weigh less than 118 lbs.

348 CARD GAMES AND MATHEMATICS

PLAYING WITH A DECK OF CARDS IS JUST A GOOD PASTIME for some people, but it could be a gamble for others. In a pack of 52 cards, there is scope for various mathematical calculations and tricks commonly used by magicians.

Mathematicians use a deck of cards for probability calculations too. Playing cards are used by children to play memory and matching games, multiplication games, pairing games, etc.

Some teachers have devised methods to teach children basic addition and subtraction using cards. Children find playing cards intriguing and fun, and therefore learn various math skills quickly. Cards also help children learn different shapes and how to segregate and sort cards based on shapes.

349 FAST FACT...

IT IS BELIEVED that a whole deck of cards represents 52 weeks of the year, and the four suits represent the four seasons!

Tarot readers and astrologers use a deck of cards for fortune telling purposes. They believe that drawing out a certain card can be an indicator of a person's future. This field is also called cartomancy. There is a system which shows trends or possibilities like achieving fame, money, sickness, etc., via the card.

Although, many people believe that cartomancy is a scientific field, there are many who criticize it heavily as being a hoax meant to misguide innocent people.

350 MATH IN CASINOS

🎓 **MATHEMATICS IS SUPPOSED TO BE VERY IMPORTANT TO CASINOS** as it is the math in their games that earns them money. It is believed that the casino always wins because of the mathematical advantage it enjoys over the players.

Casinos always take away a certain percentage of the winnings and it is considered that "by the law of large numbers," the casino always makes profit in the long run. There are several types of percentages employed in large casinos like hold percentage, win percentage, etc.

Knowledge of mathematics is necessary for the casino operator to ensure that the players' expectations are met and they keep returning to the casino frequently. Adequate math knowledge is important for the players to face the consequences of their betting activities.

Casinos come under the category of companies that are well regulated when it comes to mathematically related issues. Casino managers need to be well-versed in gaming regulation to ensure that customers get a fair deal in their wins or losses.

Many people believe that in casinos, there is nothing called luck, there is only good math and bad math.

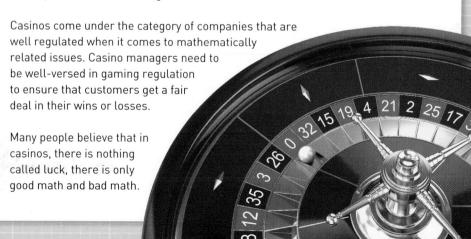

351 RELIABILITY THEORY

RELIABILITY THEORY STUDIES THE PROBABILITY OF A SYSTEM functioning successfully over a period of time. All man-made systems cannot be considered absolutely perfect and efficient. There are always glitches and imperfections because of human errors or some problem in the environment. These inherent imperfections of the system may lead to a system failure at some point in time.

Sometimes, these failures in performance may lead to an economic consequence or loss. Mathematicians have developed a theory to optimize the reliability of a system and the probability of its successful operation, like an aircraft, to ensure that the risks involved in the functioning are reduced. These predictions help improve the efficiency of the system.

Mathematicians have also developed reliability theory to assess human reliability. Based on the environment a human being works in, techniques are developed to judge how efficient the person is, and what must be done to reduce the risks that come with their imperfections.

Therefore, a careful study of industrial applications and human applications' reliability is necessary for mathematicians to develop precise models of work in the present age where even the tiniest of error may lead to grave consequences.

352 THE GAME THEORY

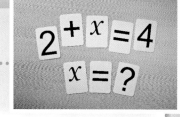

GAMES ARE PLAYED by everyone, young and old. The idea of games is to outwit your opponent.

Mathematicians have analyzed this simple idea of fun into complex mathematical models. They call such models the "Game Theory".

Using this theory, mathematicians attempt to predict the outcomes of real life situations. These predictions are made based on the possible choices players (rational decision makers) would make.

353 FAST FACT...

JOHN. F. NASH won the Nobel Prize in 1994 for his contribution to Games theory. Nash however, suffered from paranoid schizophrenia and was even institutionalized.

Mathematicians predict outcomes in different models where the decision makers have conflicting, non-conflicting or partially conflicting goals within the same system.

In 1713, the first mathematical solution of a game involving two players was written. Till the 1940s, the game theory studied situations where goals of the two players were conflicting and one player's loss had to be equal to the other player's game. These were named 'zero-sum' games like, for example, a game of poker.

A path-breaking model in games theory called the "Prisoner's dilemma" was formulated by A.W. Tucker.

In the Prisoner's dilemma the outcome depended on how the players responded jointly. It was calculated that the outcome in these games would be non-zero. In fact, both players could win at the same time—a 'win-win' situation.

In the 1990s, game theory was extended to "games" with more than two players to fit better in real life situations. Game theory has proved invaluable in diverse and immensely important fields such as environmental issues, economics, drug usage, and disarmament.

354. THE PRISONER'S DILEMMA

IN THE BEGINNING, the "games" considered were zero sum games, i.e., there could only be one winner out of two players. The winnings would always equal to what was lost.

Soon enough, models of non-zero games were developed. In this case, it was understood that the loss may not always be equal to the gain. Sometimes, both players may win.

A.W. Tucker designed a model famously called the "Prisoner's Dilemma". In this model, two prisoners who were a part of a team were picked up by the police.

The prisoners were kept separately and could not communicate in any manner. They were both told separately that if (for instance) A confessed and B did not, then A would get a very light sentence while B's sentence would be increased.

However, if B confessed and A did not, then B would get the light sentence and A the heavy one. If neither confessed, then none would be sentenced. However, if both confessed then they would both get very heavy sentences.

In this case, there are many possible win and lose situations. The outcome depends on how they both respond. There may even be a win-win situation where both remain silent and no one gets sentenced.

355 MAGIC SQUARES

IN A MAGIC SQUARE, each row and column must add up to the same number. Legend has it that a turtle with a 3x3 magic square on its back appeared out of the Luo river in around 2800 B.C. in China. The "magic" square' is based on the Latin square discussed by Euler in 1779 A.D. However, in Latin squares, the symbols did not have to be numbers and therefore did not have to add up.
A gentleman named Simon de la Loubere wrote down a method for constructing magic squares as early as the 17th century.

356 THE FIFTEEN PUZZLE

🎓 **IN THE 1870s,** Sam Loyd invented a seemingly simple game called "the Fifteen Puzzle". He offered 1,000 dollars (an extremely princely sum at that time) to the first person who could solve the puzzle. The prize was never claimed.

The puzzle consisted of 15 blocks numbered 1 to 15 enclosed in a box. They were arranged in 4 rows and 4 columns, one block remained empty. 13 blocks were arranged numerically, but block 14 and 15 were interchanged.

357 FAST FACT

📖 **IT IS POSSIBLE TO** shift the blocks in the Fifteen Puzzle to turn it into a 'magic square' with each side totaling 30.

The challenge was to shift one block at a time in order to bring these two blocks in numerical order. This puzzle became so popular in America and Europe that people were found trying to solve it even during work hours.

People became so engrossed in the puzzle that tradesmen forgot to open their shops, navigators forgot to navigate, and motor men forgot to stop at train stations.

Mathematicians soon spoiled the fun by proving that the problem posed was incapable of solution. Every place that a block was shifted from where it should be was called a disorder by the mathematicians.

Mathematicians proved that when the number of disorders was even, the blocks could be re-arranged into a numerical order. However, if it was odd, the puzzle was insoluble, i.e., it could not be brought to a regular order.

358 SUDOKU

🎓 **MAGIC SQUARES HAVE CERTAINLY WEAVED THEIR MAGIC** over millions around the world. Though the puzzle was invented in the late 1970s, its popularity only caught on once it was introduced in Japan in 1984 by Nikoli.

359 FAST FACT...

📖 **THE POPULARITY** of Sudoku or Su Doku equals that of Rubik's cube, another puzzle based on logic, which was at the height of its popularity in the 1980s.

360 FAST FACT...

📖 **POPULARITY OF** this mathematical puzzle showed up as an increase in the sales of newspapers that carried the puzzle, forcing others to carry it too.

It was Nicoli who named it "Sudoku". "Su" means "number" and "doku" means "single" in Japanese. The entire word means "the number must be single."

In Sudoku, a number can occur only once in any row or column. Nicoli still holds the trademark for this name. The Times began publishing it in 2005 under the name Su Doku.

People who are not remotely interested in mathematics can also be seen hunched over newspapers trying to solve puzzles based on these squares.

In the most popular version, the challenge is to fill out nine 3x3 magic squares placed in a 9x9 grid using numbers 1 to 9 just once in each row and each column.

361 LOTTERY

🎓 **SOME PEOPLE USE PAST STATISTICS AND COMPUTERS** to predict lottery numbers. Surprisingly, such people are successful at this art. The use of probability theory is important to make predictions like lottery results.

People master the ability to judge the numbers to be drawn by studying the samples of past lottery numbers. Many critics feel that it is impossible to predict a random event such as a lottery draw; others feel that hit-and-trial methods sometimes work to predict the outcome of these events.

Lottery experts believe that such events are based on the chaos theory where seemingly random events may actually be loosely connected. Many experts have designed tables to help them reduce the randomness of lottery draws in order to predict results with better accuracy.

The most important fact to be considered is that predicting random numbers using tools of mathematics is a popular pastime for people who buy lottery tickets in large numbers.

362 FAST FACT...

📖 **BRITISH ILLUSIONIST** Derren Brown correctly predicted all the numbers of a British lottery live on television, leaving viewers surprised and puzzled!

363 FAST FACT...

📖 **PEOPLE ARE KNOWN** to hide their lottery tickets in socks, freezers, and even taped to their chests!

364 FAST FACT...

📖 **A LOTTERY WINNER** danced so hard with joy that he broke his leg!

365 FAST FACT...

📖 **THE LONGEST CELEBRATION** for winning a lottery was at a pub for two weeks!

366 TERROR ATTACKS

WAR AND TERROR attacks seem to be very random events. But experts believe that it is not so and these can be predicted using some logic and math skills. It is believed that if carefully analyzed, these insurgent attacks show a definite path or direction, making them possible to predict. A lot of factors have to be taken into account, like the political scenario, economy, etc.

Some experts studying attacks in countries like Afghanistan and Iraq have concluded that it is possible to frame a mathematical equation in order to figure out the next possible attack and the intensity of its impact.

The team doing this analysis used publicly available information to plot previous events and develop an estimated graph to predict a future possible attack.

The most important factor in developing an equation like this is to understand the psyche of the prevalent insurgent groups. The University of Maryland developed a software called SCARE (Spatio-Cultural Abductive Reasoning Engine) that can predict the locations of guerrillas' bomb caches almost up to half a mile!

367 FAST FACT...

📖 **ON SEPTEMBER 17, 2001,** commuters used ferry services for the first time since 1883 between Brooklyn and Manhattan to reach to their workplaces, which were affected by the World Trade Center disaster.

368 FAST FACT...

📖 **NORTH AMERICA SUFFERS** fewer attacks as compared to the rest of the world.

369 FAST FACT...

📖 **THE ODDS OF** a person dying in a terror attack are far lower than them dying of any other cause.

370 FAST FACT...

📖 **THE SCARE** software can also consider cultural factors such as territorial allegiances when predicting locations.

371 FAST FACT...

📖 **INSURGENCY IS A TOOL** used to attack and reduce public confidence.

372 ECONOMIC RECESSION

🎓 **ECONOMISTS AND FINANCIAL EXPERTS USE MATHEMATICAL TOOLS** specially designed for measuring the progress of the economy. These tools help analysts predict the arrival of periods of boom or recession in the economy.

Economic experts track the financial system, study, and record all the changes happening on a daily basis. Their close monitoring of economic policies helps them foresee any difficulties that might be faced by the country in the near or long term future.

Several financial analysts make models based on many years of study while determining the dynamics of the economy and working of financial systems. The advantage of using mathematical models for predicting financial crisis is imperative in these times.

A financial crisis creates huge losses and damages economic systems across the globe making some countries bankrupt. It has a huge impact on the peace and harmony of a country, leading to large scale unrest. By predicting a looming crisis, economies can construct bail out plans and restore equilibrium in the country.

373 FAST FACT...

📖 **ECONOPHYSICS IS A NEW AREA** of study where physicists study the attributes of a country's economy to analyze the current scenario and give guidance for the future.

374 FAST FACT...

📖 **77% AMERICANS** lived "pay check to pay check" in 2012-13.

375 FAST FACT...

📖 **A STUDY REVEALS** that if one starts with a dollar and doubles the amount everyday, they can own a huge chunk of the planet's assets.

376 FAST FACT...

📖 **THE COMPANY APPLE** has cash and investments equal to the GDP of Hungary, and more than those of Vietnam and Iraq put together.

377 FAST FACT...

📖 **SINCE THE RECESSION** of 2009, more than $ 8 trillion of lost wealth has been recovered.

378 HOW MUCH MONEY WILL YOU MAKE?

ANNUITY CALCULATION IS A METHOD OF ACCUMULATING a lump sum of money through a series of payments that are equal and regular. Mathematicians use the concept of compound interest in which interaction of value, time, and interest rate are calculated to determine the annuities. For example, if a person earns $100 every month for five years at 5% per annum, it calculates how much he would have earned all together.

Calculations of this kind help to understand how much money one can make and how much a person needs to set aside for loan repayments, children's education, etc. These methods are also useful in calculating how much interest a person will earn on his accumulated savings.

Mathematicians and accountants guide people in planning for their future using these simple tools. Annuity tables offer a quick way to calculate the amounts for people who have no math training or skills.

379 FAST FACT...

AN ANNUITY THAT has no end is called a perpetuity.

380 FAST FACT...

ANNUITY CALCULATIONS are also called Time Value of Money.

381 CLIMATE CHANGE AND NATURAL DISASTERS

🎓 **CLIMATE CHANGE IS AN IMPORTANT ISSUE** that we face in today's world. It is a well-known fact that the average temperature of the Earth's surface is increasing steadily. If this increase continues, it will spell doom for life on Earth. Scientists use mathematical formulas to predict the temperature increase and devise methods to stall global warming.

Even if methods are developed to stop the heating of the Earth's surface, would it mean that temperatures will become stable? Mathematicians come into the picture to determine if this scenario would be safe after all!

Mathematicians use tools to predict rainfall, snow, tides, etc. The data obtained and recorded is used to analyze past climatic trends, changes in the trends, and final picture in the long run. Tools are developed to study cyclones, tornadoes and hurricanes, and mathematical data is used to understand what causes these disasters and how can they be averted.

Devises and scales are developed to study earthquakes and measure their intensity, and predict when and where the next one would strike.

382 FAST FACT...
📖 **THE AVERAGE SEA LEVEL** is expected to rise between 7 to 23 inches before the end of this century.

383 FAST FACT...
📖 **ABOUT 1,000 GLACIERS** out of 1,100 are receding.

384 FAST FACT...
📖 **IF GLOBAL TEMPERATURES** increase at the current rate; 20% to 30% of the flora and fauna will be extinct.

385 FAST FACT...
📖 **TEMPERATURES ARE** expected to rise between 3.2° to 7.2° F in the 21st century.

386 FAST FACT...
📖 **THE CONCENTRATION OF** carbon dioxide in the atmosphere is 387 parts per million (ppm).

387 POLLS AND NATIONAL LEADERS

🎓 **STATISTICIANS CARRY OUT POLLS** and surveys to figure out which candidate will be elected as the next leader. They also carry out surveys to understand if the current national leaders are popular with the masses.

Mathematicians take large sample sizes of people and figure out their opinions and preferences. They then take an average or mean of the people's opinions and project a final outcome. This method is beneficial in giving a general picture of what the masses think.

Sometimes these predictions are precise and correct. At other times, the results may be flawed. The method of collecting sample sizes has a large influence on the final outcome. If the process is carried out scientifically, the results emerge correctly.

Many researchers use probabilistic models and aggregate poll results to achieve accurate predictions, making them very reliable to the people of the country.

388 FAST FACT...

FUZZY MATHEMATICS is a branch of mathematics that is related to the fuzzy set theory and fuzzy logic. It was first seen in the year 1965, in Lotfi Asker Zadeh's work.

389 FAST FACT...

IN ONE OF the presidential debates held in Boston, Texas, George. W. Bush used the term "fuzzy math" four times to ridicule the arguments made by Al Gore.

390 FAST FACT...

"FUZZY MATH" is a term used by politicians to describe numbers with regard to government expenditure, which they claim to be either inaccurate or doubtful.

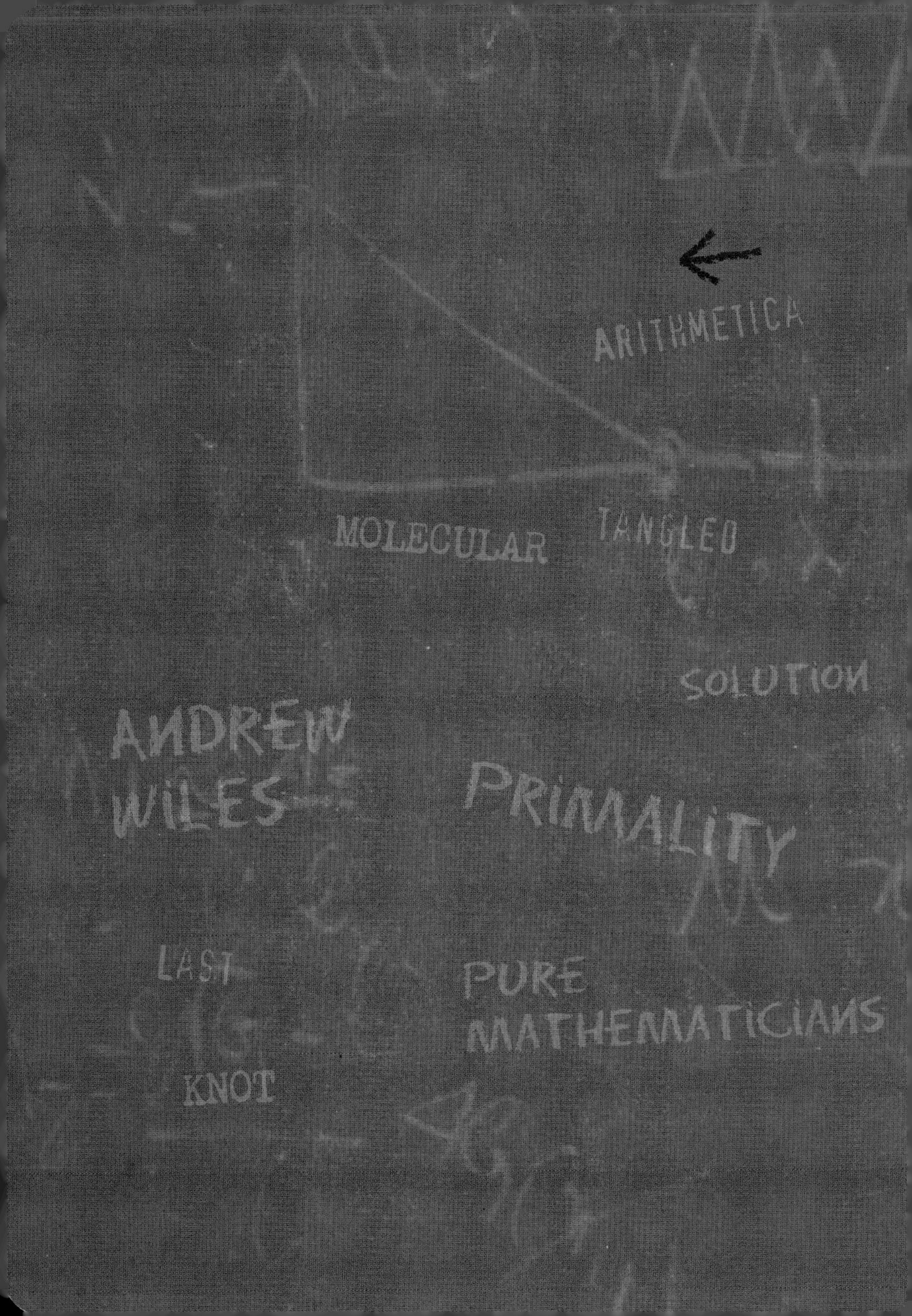

ARITHMETICA

MOLECULAR   TANGLED

SOLUTION

ANDREW
WILES   PRIMALITY

LAST   PURE
                MATHEMATICIANS

KNOT

501 Topology and Fermat's Theorem

PRIME

PURE

REPTEND

MÖBIUS

DNA

PROOF

DIOPHANTUS

391 UNTYING THE KNOT

🎓 **TOPOLOGY IS A BRANCH** of mathematics that studies shapes which are totally flexible, unlike Euclidean geometry, which is the study of rigid shapes.

Topology does not study properties such as angles and lengths like conventional branches of mathematics, but studies properties that remain unchanged by bending, twisting, or stretching.

According to topologists, a cube and a sphere are the same, and a coffee cup is no different than a doughnut. In topology, parallel lines could meet and even cross.

Knot theory is a branch of topology. It was inspired by the knots we see in everyday life, such as shoelaces and ropes.

A mathematical knot, however, differs as the ends of the knot are joined and cannot be undone. Thus, a knot is a close that does not intersect, and exists in 3-dimensional space.

The mathematical study of knots was initiated by Carl Friedrich Gauss in 1833. The earliest significant contribution to the knot theory was in Lord Kelvin's atomic theory. Lord Kelvin stated that all matter was made of ether, and atoms were knots in the ether.

Carl Friedrich Gauss

In more recent times the knot theory has been extremely useful in the study of DNA as it is a long, thin molecule that is extremely knotted and tangled.

In the last few decades, the theory has gained popularity in helping to understand underlying chemistry, molecular biology, and the world of quantum mechanics. In fact, it is most useful in the universal Theory of Everything!

To further understand the theory, the Journal of Knot Theory and Its Ramifications was founded in 1992.

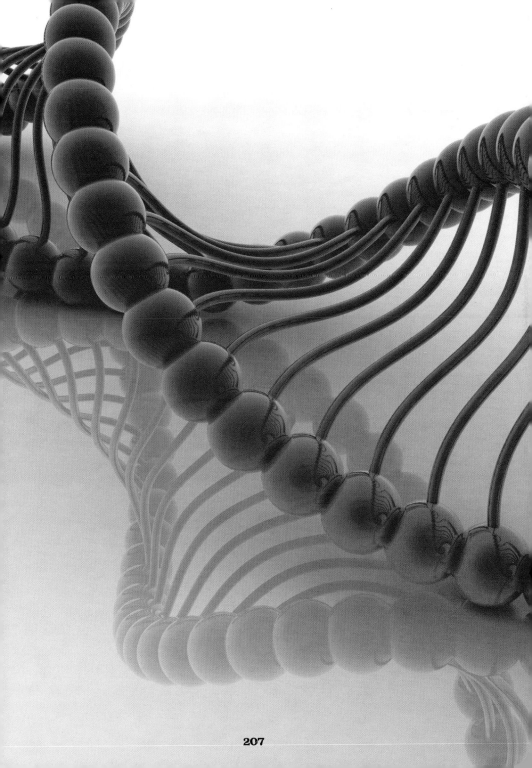

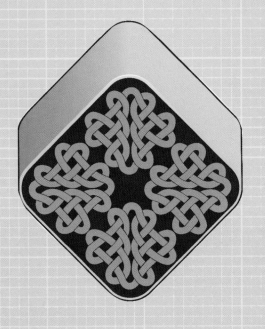

392 FAST FACT...

📖 The **KNOT THEORY IS** indispensable in the study of topographically different arrangements that have the same chemical formula but very different properties.

393 FAST FACT...

📖 **A MÖBIUS STRIP** is a strip that has just one side and one edge.

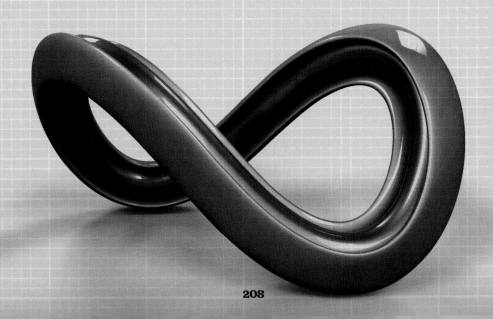

394 FERMAT'S LAST THEOREM

FERMAT CLAIMED TO HAVE DISCOVERED proof that the Diophantine equation $x^n + y^n = z^n$ had no integer solutions for n>2 and x, y, z is not equal to zero. He was quoted saying, "It is impossible for a cube to be the sum of two cubes, a fourth power to be the sum of two fourth powers, or in general for any number that is a power greater than the second to be the sum of two like powers. I have discovered a truly marvelous demonstration of this proposition that this margin is too narrow to contain."

A result of Fermat's marginal note is the proposition that the Diophantine equation have x, y, and z as integers, and have no nonzero solutions for n>2, has come to be known as Fermat's Last Theorem.

The above was called a "theorem" on the strength of Fermat's statement, despite the fact that no other mathematician was able to prove it for a long time.

The theorem was proved after hundreds of years by Wiles and R. Taylor in late 1994. This marked the end of an era in the field of mathematics.

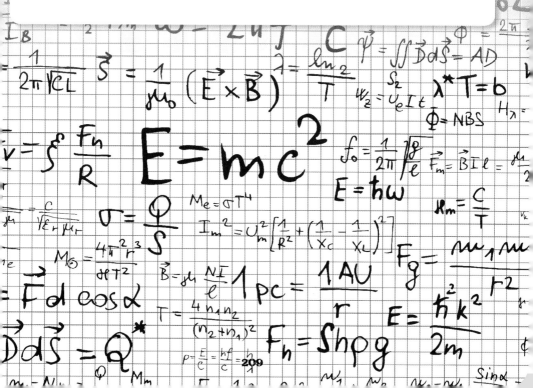

395 FAST FACT...

📖 **FERMAT HAD SCRIBBLED** the theorem in the margin of the book "Arithmetica" by the Greek mathematician Diophantus.

396 FAST FACT...

📖 **THE SCRIBBLED NOTES** of Fermat's last theorem were discovered after he died.

397 ANDREW WILES AND THE SOLUTION TO THE THEOREM

🎓 **ANDREW WILES SPENT A GOOD PART OF HIS CAREER** trying to prove Fermat's Last Theorem. He thought he had the correct solution to it in 1993, but unfortunately, some errors were found in his calculations. He was at the risk of losing his life's hard work. Sidelining his mistakes, he went back to fixing the holes in his solution and eventually came out successful.

As a child, Wiles was interested in Math. He read about Fermat's Last Theorem in a book from his library. He was intrigued by the fact that mathematicians across the world had not been able to solve this puzzle for hundreds of years!

Fermat himself could not find a comprehensive solution but never believed that there was no solution to the puzzle related to the Pythagorean equation $a^2 + b^2 = c^2$.

In mathematics, to prove something, one has to give a line by line reasoning to make sure that every aspect of the problem is covered and all the steps are self evident.

The Theorem became an obsession for pure mathematicians to prove because Fermat hinted on having the proof himself.

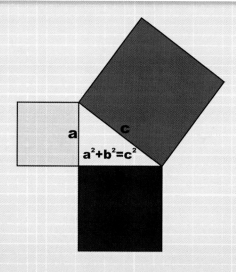

398 FAST FACT...

📖 **FERMAT SAID** he had proof. Unfortunately, all he ever wrote down was, "I have a truly marvelous demonstration of this proposition which this margin is too narrow to contain."

399 FAST FACT...

📖 **TANIYAMA-SHIMURA** conjecture was another unsolved math problem, just like Fermat's Last Theorem.

400 FERMAT PRIME

A FERMAT PRIME IS A FERMAT NUMBER that is prime. Fermat primes are near-square primes.

Fermat conjectured in 1650 that every Fermat number is prime.

In 1844, Eisenstein proposed a problem with a proof that there are an infinite number of Fermat primes.

At present, however, there are only a few Fermat numbers for which primality or compositeness has been established.

The only known Fermat primes are:
F_0 = 3
F_1 = 5
F_2 = 17
F_3 = 257
F_4 = 65537

It seems unlikely that any more will be found using current computational methods and hardware. It follows that it is prime for the special case when together with the Fermat prime indices, it gives the sequence 2, 3, 5, 17, 257, and 65537.

A Fermat is a prime if and only if the period length is equal. In other words, Fermat primes are full reptend primes.

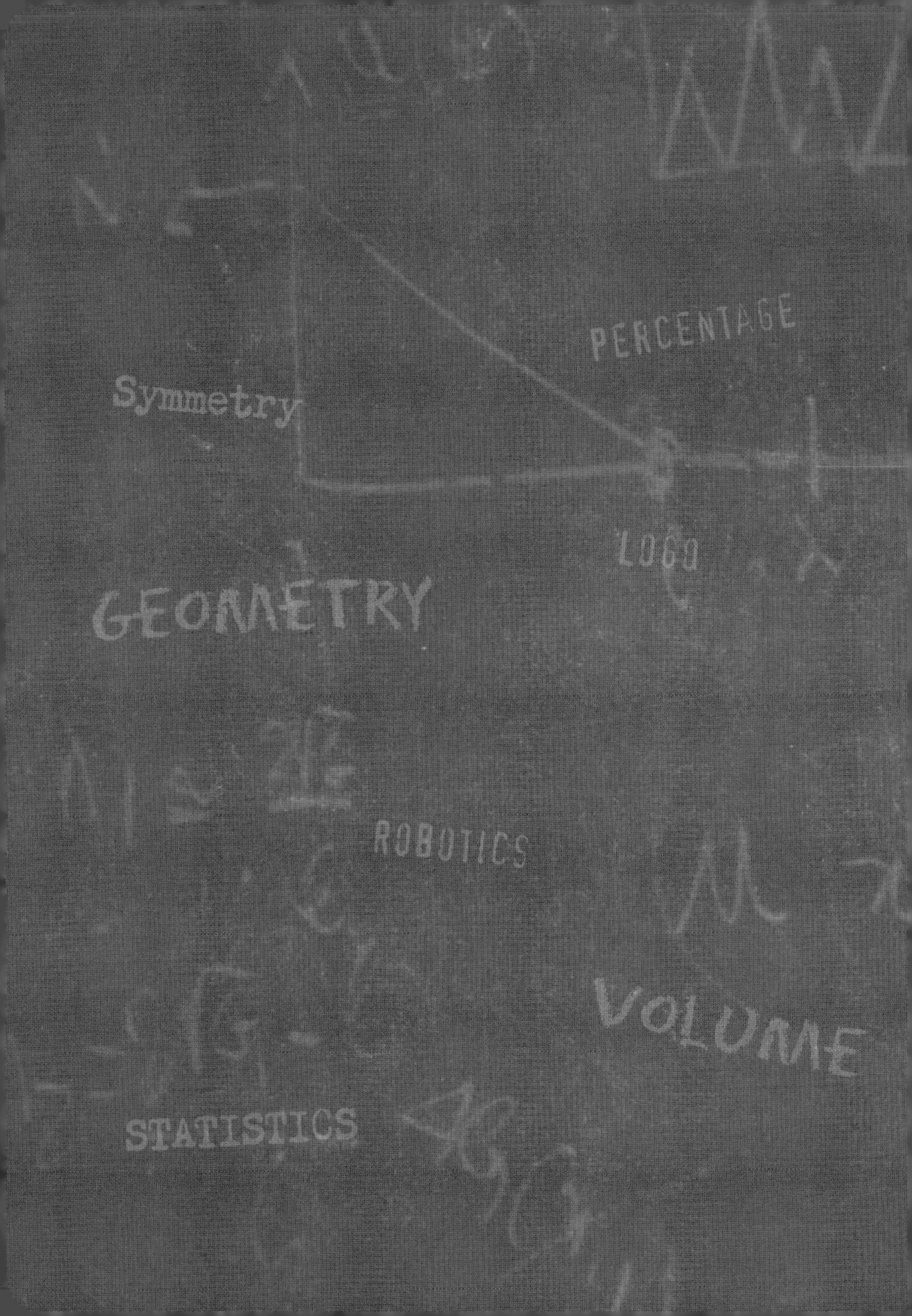

Symmetry

PERCENTAGE

LOGO

GEOMETRY

ROBOTICS

VOLUME

STATISTICS

501 Everyday Math

FOLDING

AREA

SAVING

BARCODES

ICOSAGON

KISSING NUMBER

TAXICAB

MOVIES

401 PERCENTAGE AND ITS IMPORTANCE

WE USE MATH in our daily life, sometimes without even recognizing it. We look at products in our shopping carts with some information in the form of percentage. For example, percentage of fat content in food items "12% fat, 15% sugar" means that particular product is 12 grams of fat per total of 100 grams.

Percentage denotes anything per hundred. A clear understanding of this concept helps all of us to figure out the meanings of words like "100%" which means "total".

The roots and use of percentages have been traced down to the Roman period where taxes were levied on goods per 100. Today percentages are used to denote cost of goods, discounts offered on products, interest applicable on loans, profits earned by companies, growth rate of countries, deficit or surplus projected in budgets, commission on sales, studying and comparing data, representation of data, etc.

A world without the use of mathematical percentages would be difficult to live in, considering almost everything is denoted using this method.

402 FAST FACT...

📖 **THE WORD "PERCENT"** is derived from the Latin word "per centum" meaning "by the hundred".

403 FAST FACT...

📖 **"CENTO" OF THE TERM "PER CENTUM"** began to be denoted as two small circles separated by a slanting line.

404 LOGOS AND SYMMETRY

🎓 **EVERY COMPANY** around the world has a logo to represent its brand. From car manufacturing to electronics and computers, every registered company uses some symbol to enable the consumers relate to products with the given brand.

Designers who develop these logos or symbols need to pay special attention to the pattern, because each logo conveys a message about the company it represents.

Careful use of mathematical symmetry is required to ensure that the design conveys the message that the company wants to put forth to its consumers. Since logos become the first step in marketing, the product designers make sure they develop an attractive symbol.

Some logos have hidden meanings, others have numbers, and some even have puzzles built in them. Logos are an important tool for business houses to ensure that the common man retains the image forever in his mind.

406 FAST FACT...
📖 **THE BASKIN ROBBINS LOGO** has a hidden number "31" in it to represent its 31 flavors.

405 FAST FACT...
📖 **THE HUMAN RIGHTS CAMPAIGN** logo is an equal to sign, and denotes equality for all.

407 FAST FACT...
📖 **THE CHOCOLATE COMPANY** Toblerone has a mountain as its logo. When we look closely there is a bear hidden in it, symbolizing Bern the "City of Bears." It is also the city where the chocolate was created.

408 FOLK ART AND GEOMETRY

HUMAN CREATIVITY can be traced back thousands of years. Beautiful cave paintings and inscriptions dating back many centuries demonstrate how the human mind worked in creating patterns and shapes with hand-made tools. Geometric designs are created around the world by various cultures, and each has a different style and several unique qualities.

Latvians use tiny squares to form grids, like a graph in math to create symmetrical designs. Africans use a similar method to design pots and urns. The Africans have even used the Pythagorean Theorem in their sand drawings.

The weavers of India create precise mathematical designs to craft varieties of fabric and specialized embroidery. Geometric exploration is observed to go well with folk art for creating masterpieces that can last several centuries.

409 FAST FACT...

WYCINANKI IS A Polish folk art of beautiful paper cut outs. Complex designs are created by repeating symmetrical patterns. The art pieces are used as decoration in homes.

410 FAST FACT...

THE HUICHOL MEN of Mexico used bags with repetitive geometric patterns, called Bolsa.

411 AREA AND VOLUME MEASUREMENTS

🎓 **AREA AND VOLUME** measurements are required in our daily lives for a variety of things. We calculate the area of land to know how much material would be required to build a house. We also use volume measurements to see how much paint would be needed. In the kitchen we use volume measurements to make our favorite recipes.

412 FAST FACT...

📖 **THE EGYPTIANS WERE** the ones to determine the precise areas of geometrical figures like triangles, rectangles, circles, etc., to divide up the land after the annual flooding of the Nile.

413 FAST FACT...

📖 **THE CHEVROLET SSR** Roadster Pickup is built by using geometry; its roof is a perfect sphere, doors are rectangular prisms, and several other geometrical features have been used in the making of this car.

Area measurements are done in square centimeters, and volume measurements are done in liters and gallons. Even when we need to fill gas in our cars, we measure it in gallons.

Walking through the aisle in a super market, we find different bottles, jars, boxes, and cans showing the volume and capacity of the items inside. We use these basic mathematical concepts freely without even realizing how important they are.

414 MATH AND SAVING FOR THE FUTURE

🎓 **PEOPLE WHO SAVE FOR THE FUTURE USE SMART MATHEMATICS** to determine how much cash they would need in the coming years by basing their calculation on present day prices.

Accountants use the concept of compound interest to understand how much money should be saved for a person's retirement to live comfortably.

When money is deposited in a bank as savings, the bank gives an interest expressed as percentage. Some people like to keep their savings in bank deposits, others prefer to invest these saving in high returns schemes, and some people like to invest in the stock market for quick profits. Whatever method is used, math skills are required to ensure that adequate returns are achieved on investments.

Savings are an important aspect and there is stress given right from school to educate people to become smart savers.

415 ROBOTICS

THE FIELD OF ROBOTICS IS CLOSELY CONNECTED WITH mathematics and is being increasingly used in our everyday lives. From caring for the elderly to performing surgeries in hospitals, the study of robotics has gained a lot of prominence in today's world. We use robots in our homes for cleaning, vacuuming, and even massaging our feet!

Artificial intelligence research is being directed towards enabling robots to acquire a high level of information so that they mingle freely with humans and widen the scope of their working.

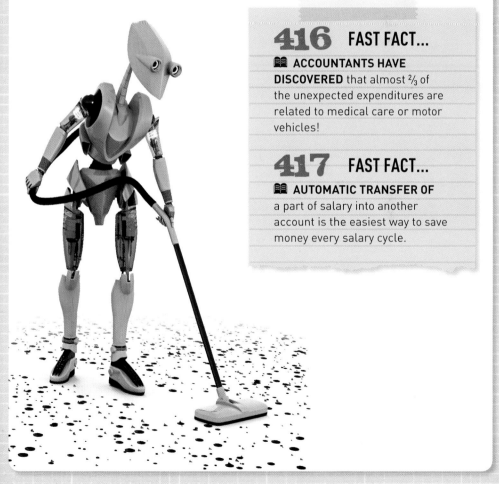

416   FAST FACT...

ACCOUNTANTS HAVE DISCOVERED that almost $2/3$ of the unexpected expenditures are related to medical care or motor vehicles!

417   FAST FACT...

AUTOMATIC TRANSFER OF a part of salary into another account is the easiest way to save money every salary cycle.

418 MATH AND DAILY STATISTICS

KEEPING A TRACK OF BASIC ECONOMIC INDICATORS is important to understand the pulse of the economy. Common people find it difficult to go through voluminous data to understand an increasingly business oriented world these days.

Mathematicians, business analysts, and accountants use computations to break down complicated figures to make them easier for people to understand.

The most important indicator is the GDP or Gross Domestic Product, which is the real market value of all the goods and services produced in the economy, indicating the society's total wealth.

The figure of money supply in the economy is necessary to understand the financial situation of the country. An index of prices is also crucial to determine inflation levels.

Numbers indicating employment levels are a must for people to understand if there is enough demand for the labor force to ensure job security among the citizens.

Information relating to housing costs and loans is necessary for people who need to build homes. Each of these statistics helps the common man to judge his environment.

419 FAST FACT...
THE STANDARD & POOR'S 500 is an index of publicly owned stocks combined into one common figure to show the overall performance of the economy.

420 FAST FACT...
NASDAQ is the US electronic stock market, where trades are done using sophisticated telecommunication and computer technology.

421 MATH IN MEDICINE

🎓 **DOCTORS AND OTHER MEDICAL CARE PROFESSIONALS** use mathematics everyday for a large number of reasons. Every prescription made by doctors involves a careful study of dosage of the medicine that must be given to the patient.

The dosage of a medicine is all about calculating the weight in milligrams of the drug that can be safely administered to the patient and the length of time it will stay in the person's body. Calculations have to be done properly or the consequences could be dangerous.

In order to prevent any mishaps, medical professionals have to be adequately trained in math skills. They follow a rigid system for writing dosaages, where they always use decimals so as to be very clear and leave no room for confusion. Doctors also need to be aware of the mathematical study of analyzing three dimensional imaging to understand CAT scans, X-Ray, and ultrasound reports of patients.

Numbers provide information to doctors about everything from blood work reports to body mass index. Numbers play a crucial role in our lives in the field of medicine.

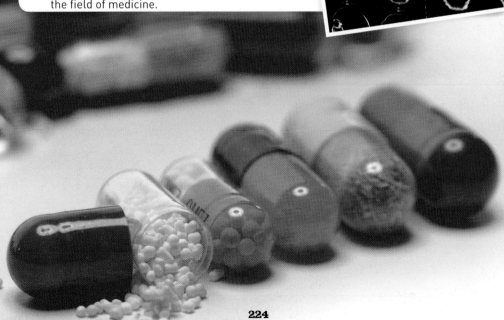

422 FAST FACT...

📖 **A LITHOTRIPTER IS** a device that uses the properties of an ellipse to treat a patient with stones in the kidney or gall bladder.

423 FAST FACT...

📖 **WHITE BLOOD CELL** counts are generally mentioned as a value between the numbers 4 to 10. However, that single value, say 8, actually means that there are 8000 white blood cells present in each drop of blood (approximately a microliter).

424 BACK TO DRAWING LINES FOR NUMBERS

🎓 **THE OPENING OF SUPERMARKETS** gave rise to the need for a system that could save time and eliminate mistakes by automatically reading product information during checkout.

Norman Woodland developed the foundations for such a system. He adapted the dots and dashes of the Morse code for this purpose. By elongating the dots and dashes used, he made vertical lines of varying thickness to be used as bar codes. This was the beginning of bar coding.

Later, optical scanners were developed. They could translate the thick and thin lines of barcodes into data such as the cost of articles instantly.

Even though the idea of bar coding was patented in the middle of the century and was used universally, it took a relatively long time to be accepted widely.

In 1966, the National Association of Food Chains (NAFC) discussed the need to automate the checkout system. A standardized 11digit code was then developed with a system to print and read the code. However, it was yet to be perfected. In 1974, a system was tried out in Marsh's Supermarket in Troy, Ohio, and was proved to be workable.

Even thereafter, manufacturers were reluctant to use the system because bar coding involved an increase in costs. Store owners, on their part, were reluctant to buy the costly scanners.

In 1977, less than 200 grocery stores in U.S.A were using scanning machines. In 1980, the number of stores using the system increased by 8,000 per year. Currently, bar coding is being used in small towns of developing countries like India.

425 THE FIRST PRODUCT TO BE BARCODED

UBCS OR UNIVERSAL BAR CODES are now found on every item in the supermarket. However, the first product to be bar coded goes to packets of Wrigley's Juicy Fruit gum. The first pack of Wrigley's gum to be picked out and scanned along with its receipt is displayed in the Smithsonian Institution in U.S.A.

Clyde Dawson was the first to pick out the pack of Wrigley's gum, and the person who scanned it was Sharon Buchanan. The famous ten-pack of Wrigley's gum was scanned at 8:01 am on 26th June, 1974, at the Marsh Supermarket in Ohio. It was sold for 67 cents.

426 FAST FACT...
WOODLAND WROTE out his first barcode in the sand on a beach.

427 FAST FACT...
WOODLAND AND HIS COLLEAGUE Silver applied for a patent on their idea of "Classifying Apparatus and Method" and the same was granted to them in 1952.

428 THE INGENUITY OF THE DABBAWALAS IN MUMBAI, INDIA

LONG BEFORE BAR CODING WAS DEVELOPED, dabbawalas (the people who deliver lunch boxes in Mumbai) have been using an extremely simple coding system.

Dabbawalas work as a team and pick up dabbas (boxes of food), from thousands of pickup points each day, and deliver each of them on time and without fail to its correct destination.

Later in the day, they take each dabba back to where it originated from. They achieve this remarkable feat based on a very simple coding system based on colors and simple numbers since many dabbawalas do not know how to read or write.

The first markings denote the collection points, the second is a color code denoting the originating station, the third is a number denoting the destination station and fourth marking informs the last dabbawala in the chain of the destination building and floor.

Dabbawalas wear a uniform of a white cotton kurta-pyjama with a white trademark Gandhi cap (topi).

The dabba system started formally as early as 1890 and has been continuously functional since then.

429 FAST FACT...

IN 2001, the dabbawalas of Mumbai were awarded the Six Sigma certification by Forbes magazine, which means that the system has an accuracy rate of 99.999999%.

430 FAST FACT...

MUMBAI DABBAWALAS featured at the Gulf Corporation Council held in Dubai in June 2013.

431 KEEPING INFORMATION SECRET

IN ORDER TO KEEP INFORMATION SECRET, it becomes necessary to code the numbers that may provide strategic information to the enemy. This enemy could be the enemy nation from whom military information is to be hidden. It could be a financial enemy from whom trade secrets have to be withheld.

The key is a system using which the numbers are coded. For example, a simple key would be that every number when replaced by one which is larger than it by 23. So, whoever has the key can decode the message.

Julius Caesar changed the letters of messages he sent through his generals using a key which was known to him and his generals only.

Television companies code their signals in such a way that only paying customers can receive the signals.

The transmission to a computer or a digital TV set involves coding the information into binary, which is a set of zeroes and ones. One of the first binary codes was the Morse code, which made use of just two symbols, the dot and the dash. It was invented by Samuel Morse who sent a message from Washington to Baltimore in 1844 via telegraph.

432 FAST FACT...
THE ROMANS ROTATED their numbers by a key to encode their messages.

433 FAST FACT...
EXTREMELY LARGE prime numbers are patented and used by banks and other financial institutions to code information.

434 FAST FACT...
THE ONLY NUMBER that has all its letters in the alphabetical order is 40 and is the only one with its letters in reverse order that are used in simple coding.

Scope for error was prevalent in this system owing to noise along the line and human error in typing. This was sought to be eliminated by making the code for each word longer, so if the encoded work made no sense, the nearest word which made sense could be looked for.

Modern code systems use code words which make detection and correction easy. Messages sent by NASA have three tiers of error correction. Even noise along the line is satisfactorily dealt with by these codes.

435 NOISE

🎓 **A NOISE IS AN ERROR WHICH IS SUPERIMPOSED** on top of a true signal. Noise may be random or systematic. It can be greatly reduced by transmitting signals digitally instead of the analog form because each piece of information is allowed only discrete values which are spaced farther apart than the contribution due to noise.

Coding theory studies how to encode information efficiently, and error-correcting codes devise methods for transmitting and reconstructing information in the presence of noise.

436 FAST FACT...
📖 **THE MATHEMATICAL** name for the division sign is "Obelus".

437 FAST FACT...
📖 **THE TERM GIGA** stands for a billion.

438 FAST FACT...
📖 **THE MULTIPLICATION SIGN** is also known as the St. Andrew's Cross.

439 FAST FACT...
📖 **IT TAKES 1,825 DAYS** for the coffee seed to yield a product that can be consumed.

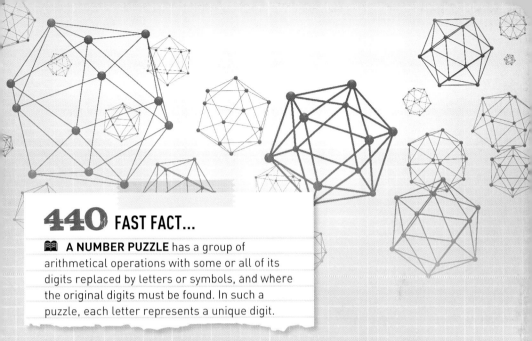

440 FAST FACT...

📖 **A NUMBER PUZZLE** has a group of arithmetical operations with some or all of its digits replaced by letters or symbols, and where the original digits must be found. In such a puzzle, each letter represents a unique digit.

441 FAST FACT...

📖 **THE SUM OF THE THREE ANGLES** of a planar triangle is always 180 degrees or Pi radians.

442 FAST FACT...

📖 **2 IS CALLED THE "ODDEST EVEN** prime number." 2 is a unique even prime because while all even numbers are divisible by 2, any number apart from 2 that is divisible by 2, is not a prime number.

443 FAST FACT...

📖 **IF YOU ADD UP** the numbers from 1 to 100, the total is 5050.

444 FAST FACT...

📖 **A PARALLELEPIPED** is a three dimensional polyhedron which is made from six parallelograms.

445 BROCARD'S PROBLEM

BROCARD'S PROBLEM asks to find the values of n for which n!+1 is a square number m2, where n! is the factorial. The only known solutions are where n=4, 5, and 7. Pairs of numbers (m, n) are called Brown numbers.

In 1906, Gérardin claimed that, if m>71, then m must have at least 20 digits. Unaware of Brocard's query, Ramanujan considered the same problem in 1913. Gupta stated that calculations of n! up to n= 63 gave no further solutions.

It is certain that there are no more solutions. In fact, Dabrowski has shown that $n! + A = k^2$ has only finitely many solutions for general A, although this result requires assumption of a weak form of the abc conjecture (if A is square).

446 FAST FACT...

IN THE MICE PROBLEM, also called the beetle problem, mice start at the corners of a regular n-gon of unit side length, each heading towards its closest neighboring mouse in a counterclockwise direction at constant speed. The mice trace out a logarithmic spiral each, meet in the center of the polygon, and travel a distance.

447 FAST FACT...

MNEMONIC IS A MENTAL deice used to aid memorization.

448 FAST FACT...

A NUMBER WITH 30 zeros is called a nonillion.

449 FAST FACT...

CHINESE POSTMEN find it problematic to create the shortest tour of a graph which visits each edge at least once.

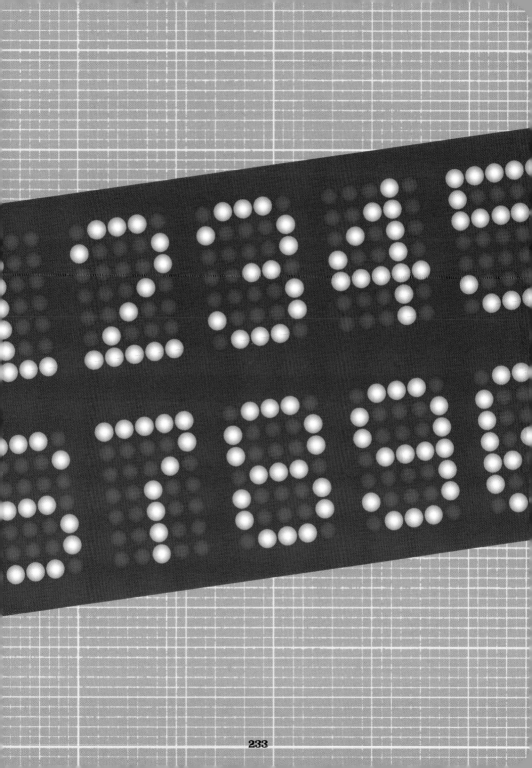

450 MRS PERKINS'S QUILT

🎓 **A MRS. PERKINS'S QUILT IS A DISSECTION OF A SQUARE** of side n into a number of smaller squares. The name "Mrs. Perkins's Quilt" comes from a problem in one of Dudeney's books, where he gives a solution for n=13. Unlike a perfect square dissection, however, the smaller squares need not be all different sizes. In addition, only prime dissections are considered so that patterns which can be dissected into lower-order squares are not permitted.

451 FAST FACT...

📖 **FROM 0 TO 1000,** the letter "A" only appears in 1000 (one thousand).

452 FAST FACT...

📖 **40 WHEN WRITTEN AS** "Forty" is the only number with letters in alphabetical order, while "one" is the only one with letters in reverse order.

453 FAST FACT...

📖 **2 AND 5 ARE** the only prime numbers that end with a 2 or a 5.

454 FAST FACT...

📖 **THE GOLDEN RATIO** of 1:618 often appears in nature and has been used to create paintings such as the Mona Lisa.

455 TITANIC PRIME

🎓 **IN THE 1980s,** Samuel Yates defined a titanic prime to be a prime number of at least 1000 decimal digits.

The smallest titanic prime is $10^{999}+7$. As of 1990, more than 1400 were known. By 1995, more than 10000 were known, and many tens of thousands are known today.

The largest prime number known as of January 2013 is the Mersenne prime $2^{57885161}-1$, which has a whopping 17425170 decimal digits.

456 FAST FACT...
📖 **SOME OF THE UNSOLVED** math problems are:
P versus NP problem
Hodge Conjecture
Riemann Hypothesis.

457 FAST FACT...
📖 **THE REULEUX TRIANGLE** is a shape of a constant width other than a circle.

458 FAST FACT...
📖 **PYTHAGORAS WAS THE FIRST** to observe that the morning star and evening star were the same.

459 FAST FACT...
NATIONAL METRIC DAY is celebrated on October 10th.

460 FOLDING

THERE ARE MANY MATHEMATICAL AND RECREATIONAL PROBLEMS related to folding. Origami, the Japanese art of paper folding, is one well-known example.

It is possible to make a surprising variety of shapes by folding a piece of paper multiple times, making one complete straight cut and then unfolding. For example, a five-pointed star can be produced after four folds, as can a polygonal swan, butterfly, and angelfish.

Amazingly, even the polygonal shape can be produced this way, as can any disconnected combination of polygonal shapes. Furthermore, algorithms for determining the patterns of folds for a given shape have been devised by Bern et al and Demaine et al.

461 FAST FACT...

THE LIFE SPAN of a taste bud is 240 hours.

462 FAST FACT...

WHEN 111,111,111 is multiplied by 111,111,111 we get 12,345,678,987,654,321.

463 FAST FACT...

WHEN THE NUMBER 1089 is multiplied by 9 the answer is 9801, the complete reverse of 1089.

464 FAST FACT...

AN ICOSAHEDRON is a geometric shape of 20 equal sides.

465 PERFECT NUMBERS

🎓 **PERFECT NUMBERS** are positive integers n such that
$N = s(n)$

Where $s(n)$ is the restricted divisor function (i.e., the sum of proper divisors of n), or equivalently $\sigma = (n) = 2n$

Where $\sigma(n)$ is the divisor function (i.e., the sum of divisors of n including n itself).
For example, the first few perfect numbers are 6, 28, 496, 8128...
Since,
6 = 1 + 2 + 3
28 = 1 + 2 + 4 + 7 + 14
496 = 1 + 2 + 4 + 8 + 16 + 31 + 62 + 124 + 248

466 FAST FACT...
📖 **21978** when multiplied by 4 we get 87912.

467 FAST FACT...
📖 **THE LARGEST** prime number so far is 12,978,189 digits long.

468 FAST FACT...
📖 **MULTIPLICATION** and succession is
1 × 8 + 1 = 9
12 × 8 + 2 = 98
123 × 8 + 3 = 987

469 FAST FACT...
📖 **A DOLLAR** can be made into small change in 293 ways.

470 NEW KIND OF SCIENCE

🎓 **A NEW KIND OF SCIENCE IS A SEMINAL WORK ON SIMPLE PROGRAMS** by Stephen Wolfram. In 1980, Wolfram's studies found unexpected behavior in a collection of simple computer experiments.

From these, he developed a methodology for tackling fundamental problems in science, from the origins of apparent randomness in physical systems, the development of complexity in biology, the ultimate scope and limitations of mathematics, the possibility of a fundamental theory of physics, the interplay between free will and determinism, and the character of intelligence in the universe.

471 FAST FACT...
📖 **AMONG ALL SHAPES** with the same area, a circle has the shortest perimeter.

472 FAST FACT...
📖 **THE OPPOSITE SIDES** of a dice cube always add up to seven.

473 FAST FACT...
📖 **ABACUS IS THE ORIGIN** of modern calculators.

474 FAST FACT...
📖 **A NUMBER IS DIVISIBLE** by three if the sum of its digits is divisible by three.

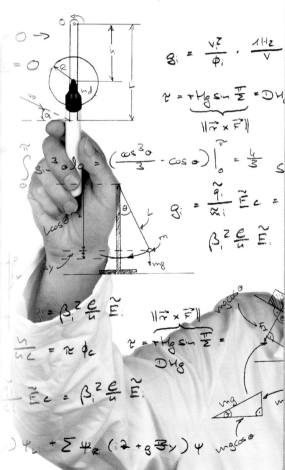

475 KISSING NUMBER

THE NUMBER of equivalent hyper spheres in ndimensions which can touch an equivalent hyper sphere without any intersections is also sometimes called the Newton number, contact number, coordination number, or ligancy.

Newton believed that the kissing number in three dimensions was 12, but the first proofs were not produced until the 19th century by Bender, Hoppe, and Günther.

More concise proofs were published by Schütte and Bartel Leendert van der Waerden and Leech. After packing 12 spheres around the central one (which can be done, for example, by arranging the spheres so that their points of tangency with the central sphere correspond to the vertices of an icosahedron), there is a significant amount of free space left, although not enough to fit a 13th sphere.

476 FAST FACT...

IN WORKING OUT mathematical equations, Greek mathematician Pythagoras used large pebbles to represent numbers. Hence the name calculus was born which means pebbles in Greek.

477 FAST FACT...

THE WORD FRACTION comes from the Latin word "Fracto" meaning "to break".

Isaac Newton

478 FAST FACT...
📖 **IN 1995 IN TAIPEI,** citizens were allowed to remove the number four from street numbers because it sounded like "death" in Chinese.

479 FAST FACT...
📖 **A "JIFFY"** is an actual unit of time and stand for 1/100th of a second.

480 FAST FACT...
📖 **IF THERE ARE 50 STUDENTS** in a class. It is almost certain that two will share the same birthday.

481 FAST FACT...
📖 **THE BILLIONTH** digit of Pi is 9.

482 MATHEMATICS IN THE MOVIES

BURKARD POLSTER AND MARTY ROSS HAVE WRITTEN A BOOK called Mathematics goes to the movies. This book is surprisingly popular. They have even made a collection of movies about mathematics and movies in which mathematics is discussed in a few scenes. It seems even non-mathematicians find it fascinating to hear Meg Ryan explain Zeno's paradox in IQ.

In Merry Andrews, Danny Kay sings about the Pythagoras theorem. In the navy Costello explains why seven multiplied by three equals 28. There are many movies which have mathematics as part of the plot.

These include Good Will Hunting, A Beautiful Mind, Stand and Deliver, Pi, Die Hard, The Mirror Has Two Faces, Proof, A Beautiful Mind.

These movies explore the Golden ratio, infinity, the other dimensions and much more. However, some movies have incorporated mathematical blunders too.

483 FAST FACT...

IN THE MOVIE 'CLOSER' based on book by Carl Sagan, it is correctly explained why use of prime numbers is a good way to communicate with aliens, though some mathematicians feel that pi is a better choice.

484 FAST FACT...

📖 **PRODUCERS WHO ARE** sincere about getting it right, hire mathematicians to make sure that they have mathematical facts right as was done for "Beautiful mind" and "Goodwill hunting".

485 FAST FACT...

📖 **IN THE WIZARD OF OZ,** the Pythagoras theorem is wrongly stated to apply to an isosceles triangle instead of a right angled one.

486 FAST FACT...

📖 **THE FILM "TWELVE ANGRY MEN"** uses the probability theory to judge the innocence of the accused.

487 FAST FACT...

📖 **TO OVERCOME THE FUEL** crisis after hurricane Sandy, New York's mayor resorted to the impartial system of allowing purchase of fuel by vehicle owners whose car number plates ended in odd numbers only on days that ended in an odd number. While, number plates that ended in zero or even numbers were allowed to purchase fuel on days that ended with even numbers. This is indeed a unique use of odd and even numbers.

488 FAST FACT...

📖 **IN MOVIES LIKE** "It's my turn" and "Straw dogs", characters were shown as mathematicians. This had no bearing to the story other than establishing the character to be one with intelligence, in contrast to movies like "Good Will Hunting", "Closer" or "Pi" where mathematics was part of the plot.

489 THE TAXICAB NUMBER 1729

🎓 **ONCE G.H.HARDY** noticed a taxi with the number plate 1729. He found it to be a rather dull number. He told his friend Ramanujan about it.

Immediately, Ramanujan corrected Hardy by saying that it certainly was not a dull number but a very interesting one.

1729 was the smallest number that could be expressed as the sum of two positive cubes in two different ways. The number 1729 came to be called the Hardy Ramanujan number.

The smallest possible natural number that can be represented as the sum of two cubes in two ways popularly came to be known as "taxicab" numbers. 1729 is the smallest nontrivial taxicab number.

In an episode of Futurama, Bender (a robot character) received a Christmas card from the machine that built him labeled, it said "Son number 1729".

The number of the Nimbus ship in episodes of Love's Labor's Lost in Space is also 1729. In Farnsworth Paradox, there is a 'Universe 1729.

490 FAST FACT...
📖 **THE SIXTH TAXICAB** number is believed to be 24153319581254312065344. However, this has not yet been proven.

491 FAST FACT...
📖 **IN THE MOVIE "BENDER'S BIG SCORE"** the taxi is numbered 87539319, which is actually the third taxicab number.

492 MATH ON THE SIMPSONS

THE TV SHOW THE SIMPSONS is the longest running sitcom. It has is famous internationally through the characters of Bart and others who are presented as simple and somewhat clumsy, they seem to talk in mathematical terms quite often.

Geometry, calculus and simple arithmetic are woven into the stories of the Simpson family. It has been suggested that a fun way to introduce mathematical topics can be through episodes of this sitcom.

In one episode, Bart tells Martin he must sit in the back row in the bus, school and at church. When questioned, he explains this is a good idea as the potential to commit mischief varies inversely with the proximity to the figure in authority.

The concept of infinity is introduced in another episode where Ned and Homer talk about who will be wearing high heels the next day. Their conversation makes the viewers wonder whether we can think about a number that is infinity plus one. In another, Bart exclaims that playing golf is a practical use of geometry.

493 FAST FACT...

IN 2012, Fox news, showed a pie chart where the percentage of people backing Romney was 60%, Palin was 70%, Huckabee 63%, thus totaling to 193%!

494 FAST FACT...

IN "THE MAN WITHOUT A FACE" the geometrical construction is wrong. A circle ABC is drawn. Then AB is joined and a perpendicular is drawn through its midpoint D. This is said to the same as DC. Since C was an arbitrary point, there is very little chance of this being true.

495 FAST FACT...

📖 **EVEN IN THE MOVIE "PI"** made in 1998, the value of pi in the opening title is wrong. Only the first 9 digits are given correctly.

496 FAST FACT...

📖 **IN THE SHOW "SERIOUS MAN"** there is an incorrect equation for standard deviation. It is said that the uncertainty in P is equal to the root of pass squared minus p squared' which would be zero! The speaker goes on to even more absurd statements.

497 FAST FACT...

📖 **A UNIQUE THEOREM** is said to exist in an episode of TV show "Futurama". When the minds and bodies of two people get tangled and cannot be interchanged, Globetrotter, a character in the show, says that the untangling can be done using two extra people. He even exclaims that "And they say pure math has no real world applications!"

498 FAST FACT...

📖 **ON DECEMBER 14,** 2011, the following "problem" was posted on Facebook. 10 + 10x0 =? People suggested using calculators, even scientific calculators, to get the answer.

499 FAST FACT...

📖 **THE MOVIE 1984'S** most iconic scene showed a man getting brainwashed into believing that 2+2=5.

500 FAST FACT...

📖 **IN A SHORT DISNEY FILM** called "Donald in Mathmagic Land" the value of pi is stated incorrectly.

501 FAST FACT...

📖 **AN ADVERTISEMENT BY** "Nature Valley" in the London Metro newspaper stated that 'we wanted to increase deliciousness by 200%, so we put two bars in each pack'! However, it seems they failed by 100% as they increased by only 100%.

INDEX

A

abacus 51, 58-9, 239
Accountants 220-1
addition 11, 19, 23, 25, 35, 50, 55, 60, 234
Africans 218
age 87, 90, 125-6, 128, 130, 132, 134, 186
algebra 30, 36, 85, 116, 121-2, 124, 135, 144, 158
   algebraic number field 124
alogon 113, 115
amount 68, 70, 170, 177, 180-1, 198, 240
Ancient Chinese 66, 106
Ancient Indians 30, 106
Angkor Wat 142, 154
angled triangle, right 21, 115
angles 8, 16, 19, 21, 30, 62, 65, 73, 108-10, 146, 206, 231
Antikythera mechanism 78
appendage 26
Arabic numbers 36
Archimedes 7, 32, 54, 71, 87, 112, 120, 126, 172
architects 11, 95, 100, 152-3, 155, 157
Aristotle 11, 32, 102
arithmetical operations 28, 30, 231
art 7, 11, 100, 150-2, 192
   folk art 218
artists 11, 95, 100, 110, 150-1, 156
Aryabhata 30, 33, 87
astrologers 10, 162, 164
astrology 7, 10, 163
astronomers 11, 76, 79, 102, 119-20, 126, 156, 158
astronomy 10, 158, 161
atoms 30, 90, 127, 206

B

Babylonians 16, 21, 31, 74, 86-7, 155, 158
banks 220, 229
bar coding 226, 228
base 17-18, 22, 29, 44-5, 48, 54, 72, 74, 85, 116, 119
beads 14, 58
beans 29, 126
Bernoulli's numbers 118
binary format 148
binary system 52, 54, 61
birds 10, 43
birth 136
Blaise Pascal 60, 106
blocks 190
bones 7, 13, 28-9, 38, 40, 45, 59
   wolf bones 45, 47
branches 202, 206
brotherhood 115, 126
Brown numbers 232
bullas 9, 14-15

C

calculations 33, 35, 48-9, 56, 58-9, 61, 75-6, 87, 114, 157, 159, 198, 211, 220, 224
   intricate calculations 147
calculators 52, 54, 60
calculus 30, 48, 52, 120-1, 124, 245
calendar 28-9, 51, 62, 72, 75-9
   calendar system 72
   day calendar 72
cards 175, 183
Carroll, Lewis 116

cartomancy   174, 183
celestial bodies   72-3, 80, 159
celestial objects   158, 161
classical dance forms   147
clay   13-15
clay tablets   15-16, 21
clay tokens   14, 17
clocks   26, 64
   mechanical clocks   63-4
code   37, 52, 74, 129, 226, 229
coins   66, 69, 97, 178
colors   111, 152, 228
COMPLEX numbers   56, 90
computers   12, 52, 58, 61, 70, 86-7, 91, 94, 107, 111, 114, 157, 161, 192, 217, 229
cones   13
construction   65, 105, 110, 157
conversion   20, 54, 67, 69
costs   216, 226
count   11, 13, 17, 40, 43-4, 57, 72, 146-7, 168, 171
countries   67, 69, 76-7, 105, 111, 120, 122-3, 158, 165, 194, 196, 201, 216, 222
credit   33
crown   120, 172
cube roots   20, 30-1, 33
cubes   9, 20, 92, 107, 122, 137, 206, 209, 244
cubit   65
cultures   41, 44, 101, 136, 143
cuneiform   9, 16, 74
curve   97, 99-100, 107
cycles   72, 78, 148

D

Da Vinci, Leonardo   98, 142, 152

dabbawalas   228
dance   143, 146-7
David Hilbert   124
death   7, 105-6, 113, 121, 125, 127-9, 131, 133-4, 136, 138, 241
Decans   62
decimals   18, 27, 50, 54, 60, 67, 146, 224
   decimal digits   235
   decimal number system   30
   decimal place-value system   33
   decimal places   21, 87-8
   decimal system   28, 30, 54, 61, 67, 72
degrees   20-1, 108-9, 161, 231
digits   41, 52-3, 60, 84, 86, 89-90, 114, 231-2, 238-9, 245
dimensions   86, 107, 116, 128, 240, 242
discoveries   52, 78, 105, 111, 122, 156, 158
diseases   134
distance   65, 67-8, 151-2, 154, 158-9, 171, 177, 232
dots   29, 226, 229

E

Earth   5, 62, 65, 67, 69, 71, 75-6, 79, 86, 108, 154, 158-60, 162, 199
eclipses   76, 78
economy   194, 196, 222
Egypt   19, 95, 119, 153
   Egyptian numbers   22
   Egyptian system of fractions   23
Egyptians   22-3, 31, 57, 62-5, 74, 219
engineers   120, 152, 156-7
English number system   92
equations   21, 34, 94, 100, 121, 135, 158, 194
equilateral triangles   105, 107

equinoxes 62, 73, 75, 153
error 49, 61, 64, 67, 75, 79-80, 111, 153, 177, 186, 211, 229-30
estimation 175, 177-9
Euclid 89-90, 95, 105, 124
Euler 84-5, 103, 189
Europe 26-7, 30, 35, 155, 190
experiment 43, 169

F

factorials 55, 85, 232
Fermat 7, 91, 130, 209-13
Fibonacci 96-7, 99, 103, 106
fingers 26, 38, 40-2, 44-5, 57, 65, 173
food 131-2, 153, 228
formations 146-7
formula 71, 85, 90, 107, 111, 118, 126, 159
fractals 9, 106-7
fractions 19, 22-3, 25, 29-31, 34, 48, 54, 87, 91, 115, 178
France 26, 55, 60, 67, 123, 136, 171
frequencies 85, 148

G

Galileo Galilei 7, 10, 55, 64, 103, 151, 158, 160
game theory 117, 187
games 91, 174, 184, 187, 190
Gauss, Carl Friedrich 90, 206
generations 61, 101, 119
geometry 43, 85, 102, 119, 124, 130, 146, 156, 158, 162, 214, 218, 245
Germany 55, 117, 124, 135
God 55, 98, 100, 139
Gödel 132

gold 66, 95, 120, 172
golden 83, 95, 145, 153
golden ratio 95-6, 98-100, 234, 242
golden rectangles 95, 99
googol 8, 93
Gothic churches 155
Gottfried Wilhelm Leibniz 52
gravity 69, 147
greater than 40, 43, 55
Greek mathematicians 21, 32, 101
Greeks 23, 32, 58, 64, 101-2, 108, 121, 160, 240
Gregorian calendar 76-7, 79
Grothendieck 133
group 62, 115, 125, 144, 146, 231

H

Hardy 113, 118, 122, 244
Harvard 123
height 65, 68, 100, 154-5, 191
heuristics 176
hexagons 110, 162
hieroglyphs 22-4
Hilbert 112, 124
Hindu-Arabic number system 30, 35
Hindu Numerals 35
Homer 40, 245
homes 60, 124, 126, 218, 221-2
hospitals 121-2, 125, 131, 136, 221
hours 19, 62, 64, 68, 74, 114, 237
hydrostatics, first law of 172
Hypatia 138
hypotenuse 21, 115

I

icosahedron 237, 240
India 27, 30, 33, 35, 54, 64, 66, 76, 91, 97, 105, 114, 118, 156, 165
   Indian mathematicians 30, 48, 97
INFINITE number 90, 91, 124, 213
infinite sets 104
infinity 8, 32, 55, 101-4, 242, 245
instruments 45, 63, 144, 148, 161
Ishango bone 39, 45-7
Isometry 143, 150

J

Jews 44, 135
Julian calendar 75, 80

K

Kasner 93
Kepler, Johannes 95-6, 160
key 9-10, 145, 229
knot theory 206, 208

L

large numbers 18-19, 29, 59, 82, 90-3, 146, 192, 224
   calculating 126
Latin squares 189
leap year 74, 76
Lebombo bone 45-7
lengths 20, 45, 65, 76, 100, 107, 145, 155, 158, 206, 224
letters 25-6, 37, 74, 85, 88, 92, 118, 122, 136, 139, 229, 231, 234
logarithm tables 59
logic 7, 11-12, 48, 93, 123, 157-8, 180, 191, 194
logos 214, 217
longitudes 108

M

machine 52, 60-1, 95, 244
magic squares 189-91
Mandela, Nelson 100
Mandelbrot set 107
maps 5, 70, 111
markings 14, 45-7, 74
Mars 79-80
masses 69, 71, 201
math skills 101, 173, 194, 220, 224
mathematical astronomy 154, 158
mathematical calculations 58-60, 72, 114, 181
mathematical concepts 145, 150
mathematical probability 178, 180-1
mathematical system 29
mathematical terms 178, 245
mathematicians 10-12, 29-30, 79, 87-8, 101-5, 111, 116-18, 124-6, 134, 148-51, 156-9, 186-7, 190, 198-9, 242-3
   ancient mathematicians 18, 20, 31, 126
   applied mathematicians 12
   pure mathematicians 12, 88, 137, 211
mathematics
   applied mathematics 12, 118
   pure mathematics 12
mathematikós 9, 11
Mayan mathematical system 29, 72

Mayans 29, 72-4, 158
measurements 65, 68, 70, 74, 109, 151, 154
memory 49, 114, 173, 183
Merkhets 63
Mersenne 82, 89, 91, 235
Mesopotamia 9, 13-14, 16-17, 19
metal 66, 172
metric system 67, 109
minutes 19, 74, 144, 167
models 12, 187-8, 196
money 20, 61, 84, 135, 167, 177, 181, 183-4, 198, 220-1
moon 10, 72, 76, 80, 160, 162
Moscow Mathematical Society 131
multiples 18, 22, 25, 90-1, 147
multiplication 19, 25, 30, 34, 55, 60, 238
   multiplication tables 20, 48, 116
music 10, 115, 142, 144-9

N

NAFC (National Association of Food Chains) 226
napier 51, 59
Napoleon 105, 136
nature 32, 62, 68, 97, 99, 107, 110, 136, 148, 152, 234
Neumann 113, 117
Nicoli 191
nobleman 126, 171
notches 45-6
notes 66, 121, 144-5
nothingness 32, 116
number plates 140, 243-4
number theory 10, 118, 122, 144
numbers

abstract numbers 15
amicable numbers 94
cardinal numbers 147
divine numbers 140
golden numbers 95
holy numbers 139
imaginary numbers 56, 84
irrational numbers 85, 91, 103
limited numbers 26
lucky numbers 136, 140
magic numbers 92, 163
negative numbers 34, 56
ordinal numbers 104
perfect numbers 82, 89, 115, 119, 238
positive numbers 34, 56
primary numbers 118
rational numbers 86, 91
real numbers 53, 56, 104
square numbers 86, 232
successive numbers 41, 96-7
total numbers 101, 146, 178
transcendental numbers 84
transfinite numbers 104
numbers rule 115
numerals 18, 20, 26, 29-30, 35
   numerical order 190
numerologists 136-7

O

Obelus 50, 55, 92, 230
Optical illusion 110

P

paper 52, 106, 117-18, 123, 150, 152, 237
parabola 86

parametric modeling   157
Pascal   60, 106
patterns   10, 69, 86, 90, 106-7, 146-8, 150-1, 217-18, 234, 237
Pendulum clocks   64
Perkins's Quilt   234
phi   11, 83-4, 153
Phidias   11, 95, 97
philosophers   10, 32, 119-20, 126, 134
philosophy   126-7, 138
pi   83-4, 87-8, 126, 153, 241-3, 245
  value of pi   30, 87-8, 92, 245, 247
pictograms   15, 22
planets   79, 158, 160-1, 164-5
Plato   10
polygons   110, 119, 232, 237
Pope Gregory XIII   75-6
position   18, 20, 31, 54, 106, 108, 156, 161, 164-5
power   30, 33, 59, 65, 84, 87, 90-3, 115, 136, 140, 209
Prisoner's Dilemma   187-8
probability   85, 117, 175, 177-8, 186
problem   12, 30, 34, 69, 74, 85, 87, 93, 98, 111, 117, 137, 155, 173, 186
process   86, 95, 99, 110, 114, 119, 153, 169-70, 176-7, 181, 201
project   61, 121, 123, 201
Pythagoras   7, 10-11, 105, 112, 115, 119, 126, 153, 236
Pythagoras theorem   21, 95, 115, 119, 155, 218, 242-3
Pythagoreans   86, 89-90, 115, 126

Q

quantities   11, 15, 17, 31-2, 45, 54, 56, 65, 91, 95, 177

R

Ramanujan   113, 118, 122, 232, 244
ratio   11, 21, 83-4, 87, 95-6, 98, 100, 145, 178, 181
rectangles   95, 99, 105, 219
reliability   175, 186
Rhesus monkeys   43
Roman number system   25-6
Romans   9, 25-6, 58, 64-5, 120, 126
roots   21, 36, 55, 68, 83-4, 114-15, 123, 126, 216, 245
rules   11, 33-4, 55, 100, 103, 108, 126, 167

S

Samoa   119
Saturn   71, 80
scale   72, 145, 159, 199
Schur   135
scientists   7, 11, 46, 76, 78-9, 102, 110, 127, 149, 159, 164, 169, 199
sculptures   95, 97, 100
sea   5, 105, 128, 161
seasons   62, 75-6, 183
sequence   96-7, 99, 213
set   30, 43, 45, 48, 52, 59, 78, 92, 96, 104, 111, 125, 144, 146-8, 177
sexagesimal system   18-19, 31, 74
shunya   34, 50, 54
signals   42, 148, 229-30
Sirotta, Milton   93
Sissa   91

software   52, 194
solar system   78, 160
solutions   12, 111, 117, 121-2, 172, 190, 204, 211, 232, 234
sphere   13, 71, 108-11, 126, 157, 206, 240
square roots   21, 82, 86
stars   31, 63, 108, 148, 159-61, 165
structures   148, 154-5
subtraction   19, 23, 25, 55, 57, 60, 183
Sudoku   174, 191
sum   34, 57, 86, 89-92, 94-7, 106, 109, 122, 137, 190, 209, 231, 238-9, 244
Sumerians   16-17, 19-20, 31, 66
Sun   62, 65, 72, 75-6, 79-80, 154, 156, 162
surface   14, 69, 108, 157
surface area   157, 159
symbols   15, 18, 22-3, 25-6, 28-9, 31-3, 40, 52, 54-5, 86, 103, 115, 119, 139, 217
symmetry   106, 146, 150, 156, 214, 217
Syracuse   120, 126
system   22, 25, 27-9, 31, 35, 40, 49, 52-4, 59, 69, 74, 180-1, 186-7, 226, 228-9
    coding system   228
    financial system   196

T

tablets   15, 19-21
teachers   20, 49, 98, 127, 146, 148, 183
telescope   158, 161
temperatures   67, 159, 199-200
theorem   12, 101, 105, 118-19, 122, 130, 132, 135, 209-11
theory   107, 127, 130, 160, 176, 186-7, 206
    geocentric   160
tokens   13-15, 17
tools   11, 46, 51, 54, 58, 144, 148, 177, 195, 198-9, 218

topology   128, 205-6
triangles   10, 83, 86, 103, 105-10, 119, 153, 162, 219
trigonometry   30, 48, 76, 108, 153, 158, 160

U

units   23, 58, 65, 67-9, 84, 99, 109-10, 241

V

value   18, 20-1, 24-5, 29, 31, 33-5, 52, 54, 57, 59, 85, 87, 89, 91, 177-8
Vedas   30, 39, 48
Vedic Math   48-9
Vitruvian Man   98
volume   49, 66, 71, 107, 121, 126, 157, 160, 214, 219

W

water   63, 71, 110, 128, 162
weight   20, 66-7, 69, 147, 153, 172, 182, 224
Weiner   112, 123
Western Europeans   41
Wiles, Andrew   204, 211
winnings   184, 188, 193
women   121, 138, 173
writing   14-16, 22, 28, 37, 60, 74, 93, 106, 139-40
writing numbers   15, 25-6, 35

Y

zeroes   93, 229
zodiac   79, 143, 162, 165